NUCLEAR EXPERIMENTS USING A
GEIGER COUNTER

by John Iovine

International Standard Book Number: 978-1-62385-008-1

Editor: Melissa Serao

Cover Concept: John Iovine

Photographs: John Iovine

Graphics: Christopher Liberti

Graphics Conversion: Christopher Liberti

Illustrations and Other Materials: Courtesy of the Author

Trademark Acknowledgements:

NUCLEAR EXPERIMENTS USING A

GEIGER COUNTER

Table of Contents

Chapter 1

1 - What is Radiation?

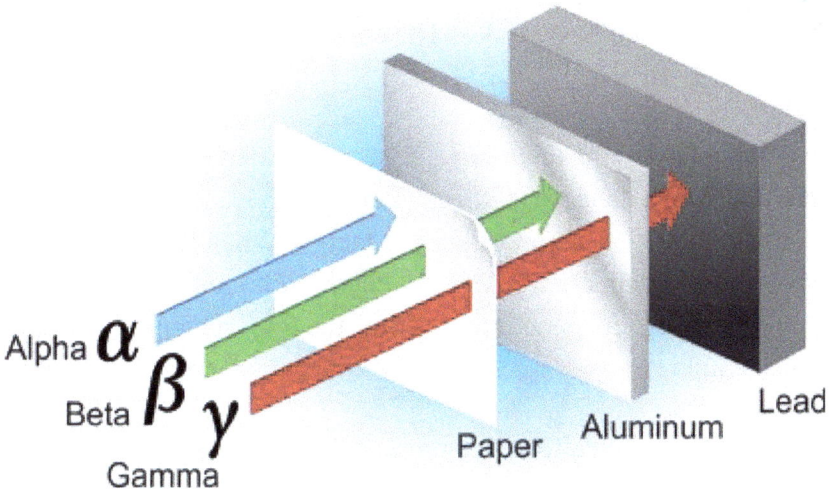

Figure 1.1

Ionizing Radiation

Before we use our Geiger counter to detect and measure radiation, we ought to define what we mean by radiation.

Electromagnetic radiation includes long radio waves to ultra-short gamma rays. In this broad view, radiation can include the heat given off from a candle, light emitted by an LED, or from particles emitted by uranium ore. To narrow our definition of radiation for the purposes here, we are confining it to what is considered ionizing radiation. Ionizing radiation is radiation that can strip electrons from atoms or molecules, thereby making the resulting atom or molecule an ion.

The types of ionizing radiation that we will be measuring and performing experiments with are gamma rays, x-rays, and beta and alpha particles.

Alpha particles are massive particles consisting of two neutrons and two protons. They are equivalent to the nucleus of a helium atom. Alpha particles have a net positive charge. This radiation has a low penetration power. A few inches of air or a piece of paper can effectively block alpha particles. The outer skin of our body protects us from alpha particles.

Surprisingly, if an alpha particle emitter is ingested or inhaled, the alpha radiation is dangerous. (See "Q" factor) Because alpha particles are massive, they have a lot of kinetic energy. When it strikes inside our body, it can damage DNA, break chemical bonds, and create tissue damage. So, while its penetration is low, the local damage, if ingested, is considerable.

Beta particles are electrons, identical to the electrons found in atoms. Beta particles have a net negative charge. They have low penetrating power. Most beta radiation can be blocked by 1/8" (4mm) of aluminum.

Gamma and X-rays are high energy ultrashort electromagnetic radiation. This classification of radiation has the greatest penetrating power. High energy gamma radiation is able to pass through several centimeters of lead and still be detected on the other side. Gamma radiation is attenuated by dense materials such as lead. Gamma radiation is produced naturally from the decay of some radioactive materials. X-rays, on the other hand, is man-made radiation used in medicine and dentistry. X-rays' penetrating power is used to take internal x-ray photographs of the human body and teeth. While X-rays are manmade electromagnetic radiation, their frequency is so high that the radiation is ionizing.

Detecting Radioactivity

There are many instruments available for detecting and measuring radioactivity. We are focused on using the Geiger counter for our detection and experiments.

Geiger counters are instruments that can detect and measure radioactivity using a Geiger Mueller tube. Geiger Mueller tubes are commonly referred to as a GM tube. The original design of the GM tube, created by Hans Geiger and E.W. Mueller in 1928, hasn't changed much, and the tube's sensor function remains the same.

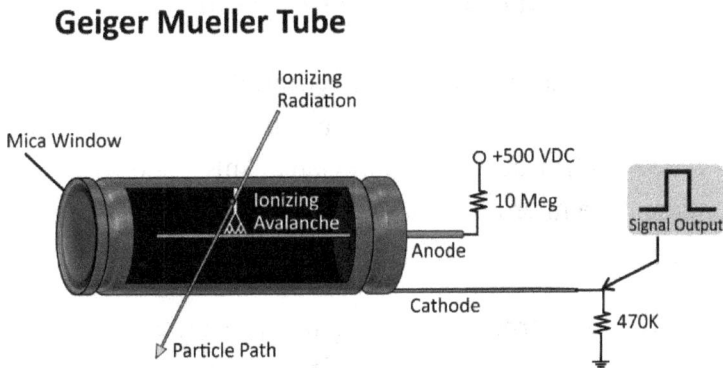

Geiger Mueller Tube

Figure 1.2

Radiation, as it passes through the GM tube, ionizes the gas within the tube. The ionization initiates a momentary avalanche of electrons accelerated by the high voltage potential used to power the GM tube. This avalanche creates a momentary conductive path between the wire at the center of the tube (Anode) and the wall of the tube (Cathode), see figure 1 resulting in a 'click' sound. By measuring the number of the clicks, the instrument indicates the radiation levels.

The GM tube sensor is the heart of the Geiger counter, and to be a Geiger counter, the device has to contain a Geiger Mueller (GM) tube. The GM Tube can detect gamma and X-ray radiation, beta particles, and if manufactured with a thin mica window, alpha particles as well.

Measurement of Radiation

Radiation Penetration Calculator

Images SI, Inc. offers a radiation penetration calculator on their website. It can be found at: https://www.imagesco.com/geiger/cal/

There are a few scales that one can use to measure radiation. Depending on your application, one scale may be better than the others.

Radiation Measurements

Roentgen: Is the measurement of energy produced by Gamma or X-Ray radiation in a cubic centimeter of air. It is abbreviated with the capital "R". One milliroentgen, abbreviated "mR", is one-thousandth of a roentgen. One micro roentgen, abbreviated "uR" is one-millionth of a roentgen.

RAD: Radiation Absorbed Dose. Original measuring unit for expressing the absorption of all types of ionizing radiation (alpha, beta, gamma, neutrons, etc) into any medium. One rad is equivalent to the absorption of 100 ergs of energy per gram of absorbing tissue.

REM: Roentgen Equivalent Man is a measurement that correlates the dose of any radiation to the biological effect of that radiation. Since not all radiation has the same biological effect, the dosage is multiplied by a "quality factor" (Q). For example, a person receiving a dosage of gamma radiation will suffer much less damage than a person receiving the same dosage from alpha particles, by a factor of three. So alpha particles will cause three times more damage than gamma rays. Therefore, alpha radiation has a quality factor of three. Following is the Q factor for a few radiation types.

Radiation:	Quality Factor (Q)
Beta, Gamma and X-rays	1
Thermal Neutrons	3
Fast n, a, and protons	10
Heavy and recoil nuclei	20

The difference between the rad and rem is that the rad is a measurement of the radiation absorbed by the material or tissue. The rem is a measurement of the biological effect of that absorbed radiation.

For general purposes, most physicists agree that the Roentgen, Rad, and Rem may be considered equivalent.

System International (SI) of Units

The System International of unit for radiation measurements is now the official system of measurements. This system uses the "gray" (Gy) and "sivert" (Sv) for absorbed dose and equivalent dose respectively.

The conversion from one system to another is simple:

1 Sv = 100 rem	1 rem = .01 Sv
1 mSv = 100 mR (mrem)	1 mR = .01 mSv
1 Gy = 100 rad	1 rad = .01 Gy
1mGy = 100 mrad	1 mrad = .01 mGy

To get an idea of this visually, try using the Radiation Dose Chart below.

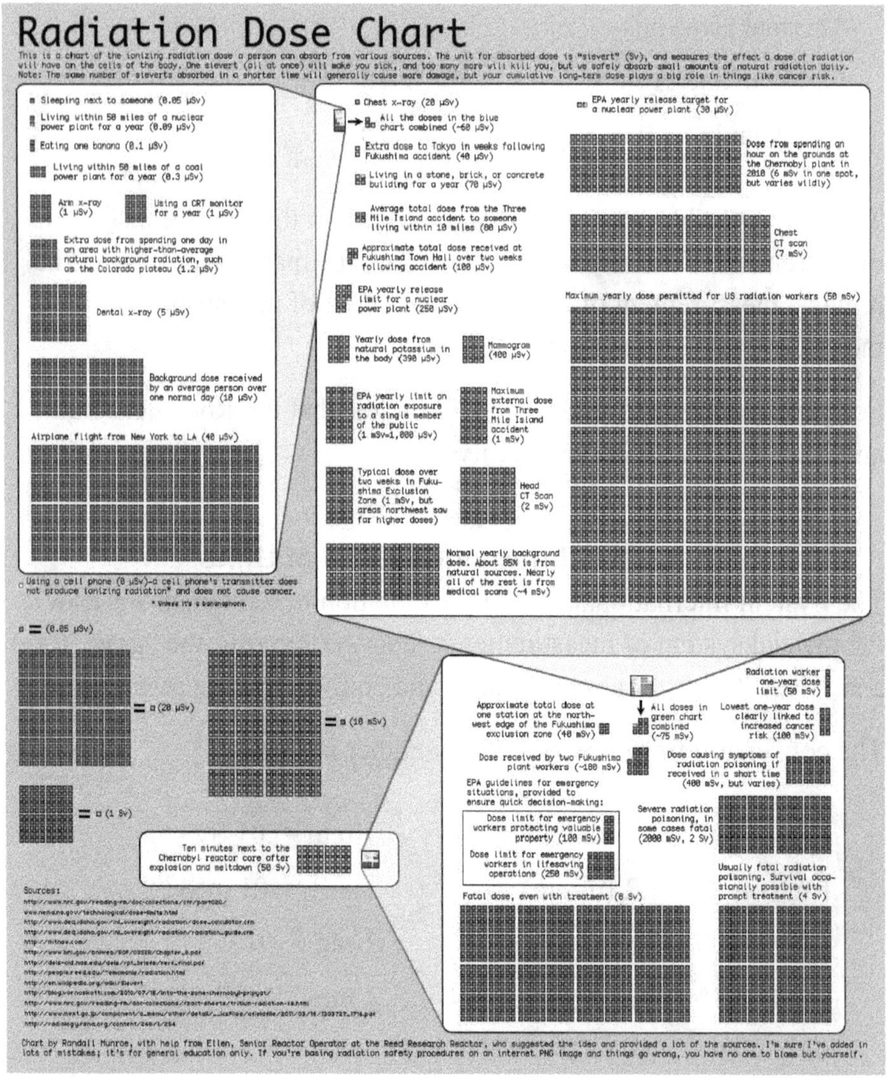

Figure 1.3

Chart in figure 1.3 is public domain. No copyright is implied. Anyone is free to use this chart anywhere with no permission necessary. Please note – This chart is for general information only. Accuracy of information in the chart is not confirmed.

How Much Radiation is Safe?

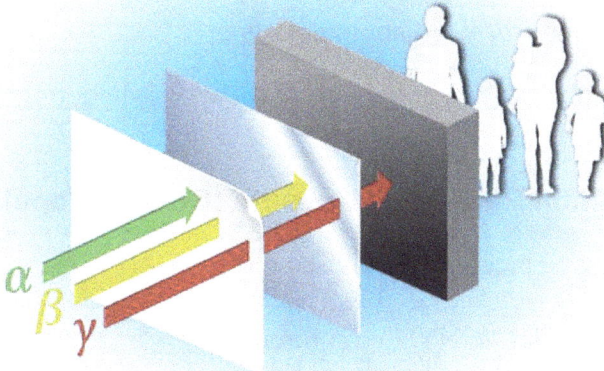

Figure 1.4

In the United States, the U.S. Nuclear Regulatory Commission (NRC) determines what radiation exposure level is considered safe. Occupational exposure for workers is limited to 5000 mrem per year. For the general population, the exposure is 500 mrem above background radiation in any one year. However, for long term, multi-year exposure, 100 mrem above background radiation is the limit set per year.

Let's extrapolate the 100 mrem number to an hourly radiation exposure rate. There are 365 days/yr x 24 hr/day equals 8760 hours. Divide 100 mrem by 8760 hours equals .0114 rem/hr or 11.4/hr millirem. This is a low radiation level. The background radiation in my lab hovers around 20 uR/hr. Am I in trouble? No. Typically, background radiation in the United States averages 300 mrem/yr. My yearly radiation exposure from 20 uR/hr is about 175 millirem/year.

Notice that my lab readings are in microrad (uR/hr) and the exposure limit is given in microrem (urem/hr). I do not know what type of radiation (a, b or y) the Geiger counter is reading in my lab at any particular instance, so I do not know the Q factor of the radiation, and

9

therefore, cannot calculate the mrem. However, for general purposes, I consider them equivalent.

The US Government's EPA has an online calculator to help you determine your yearly radiation exposure:

https://www.epa.gov/radiation/calculate-your-radiation-dose

Common Radiation Exposure (General Population)

Exposure Source	Dose(conventional)	Dose (SI)
Flight from LA to NY	1.5 mrem	.015 mSv
Dental X-ray	9 mrem	.09 mSv
Chest X-ray	10 mrem	0.1 mSv
Mammogram	70 mrem	0.7 mSv
Background Radiation	620 mrem/year	6.2 mSv/year

Background radiation consists of three sources; **Cosmic** radiation from the sun and stars; **Terrestrial** radiation from low levels of uranium, thorium, and their decay products in the soil, air, and water; **Internal** radiation from radioactive potassium-40, carbon-14, lead-210, and other isotopes found inside our bodies.

Figure 1.5

Radiation Safety - Lead Shielding Guide

Purchase Lead Sheet(s) for Gamma Shielding:

Shielding reduces the intensity of radiation depending on the thickness. This is an exponential relationship with gradually diminishing effect as equal slices of shielding material are added. A quantity known as the halving-thicknesses is used to calculate this. Halving thickness is relative to the energy level of the gamma radiation. Higher intensity radiation will require thicker shielding.

For instance, the gamma radiation emitted from Cobalt-60 are 1.33 and 1.17 MeV. If we look at the gamma radiation emitted from Iridium-192, they are 0.31, 0.47, and 0.60 MeV. The gamma radiation from Cobalt-60 has twice the energy as the gamma radiation from Iridium-192, therefore, Cobalt-60 halving thickness will be greater than the halving thickness of Iridium-192.

The halving thickness of lead is 1 cm. This means the intensity of gamma radiation will reduce by 50% by passing through 1 cm of lead.

For example:

1) A lead shield 2.0 cm thick reduces gamma rays to 1/4 of their original intensity. (1/2 multiplied by itself two times)

11

2) 3.0 cm of lead reduces gamma radiation to 1/8 of their original intensity (1/2 multiplied by itself three times)

Halving thicknesses of some materials, that reduce gamma ray intensity by 50% (1/2) include:[5]

Material	Halving Thickness[cm]	Halving Thickness[inches]	Density [g/cm³]	Halving Mass [g/cm²]
Lead	1.0	0.4	11.3	12
Steel	2.5	0.99	7.86	20
Concrete	6.1	2.4	3.33	20
Packed soil	9.1	3.6	1.99	18
Water	18	7.2	1.00	18

The Halving Mass column in the chart above indicates mass of material, required to cut radiation by 50%, in grams per square centimeter of protected area.

Calculating Radiation Halving Thickness

Images SI Inc. has a free online radiation penetration calculator located on its website at:

https://www.imagesco.com/geiger/cal/

Using this online calculator, you can estimate the halving thickness value of various materials and radiation types. For example, I put in the high gamma radiation from Iridium-192, which is 0.60 MeV. In the calculator, this is 600 keV.

Radiation Type:	Gamma ▾	
Energy	600	KeV
Medium	Lead ▾	
Intensity Decrease Factor	2	
Attenuation Percentage	50	%
	Calculate	

Output

Penetration in medium : 0.4900 Cms

Gamma - Lead

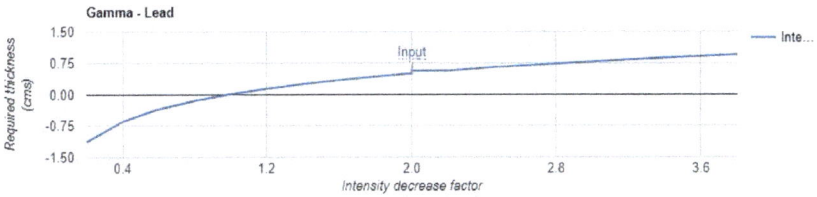

Figure 1.6

The calculator estimated the halving thickness of lead at 0.49 cm thick.

I next put in the high gamma radiation of Cobalt-60, 1.33 MeV (or 1330 keV). The calculator estimated the halving thickness at 1.08 cm of lead.

Storage of Radioactive Isotopes

Many of the experiments in this book use license exempt radioactive sources. These are 1" in diameter, 1/8" inch thick plastic discs with a small amount of radioactive isotope embedded within the center of the plastic.

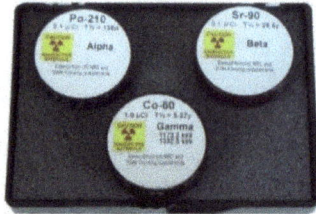

Figure 1.7

The radioactivity of these sources is so low that the United States government allows the general public to purchase and own these sources without needing a license.

Even so, to reduce your exposure to any additional radiation, you may want to purchase a lead container to hold these disc sources. These containers are called lead "pigs".

Figure 1.8

You can purchase new and refurbished lead pigs here:

https://www.imagesco.com/geiger/containers.html

Other Helpful Links

U.S. Nuclear Regulatory Commission

http://www.nrc.gov/

CDC - Center for Disease Control maintains a radiation emergency web site:
http://www.bt.cdc.gov/radiation/index.asp#clinicians

Health Physics Society
http://www.hps.org

U.S. Environmental Protection Agency
http://www.epa.gov/radiation

Author and Publisher disclaimer: We do not make any warranties (express or implied) about the radiation information provided here for your particular use. All information should be confirmed and verified with local and national government organizations or recognized experts in this field before being used.

Chapter 2

2 - Geiger Counter Buyer's Guide

If you haven't purchased a Geiger counter, this guide will help you make a selection. With so many Geiger counters for sale online, it's easy to become confused when comparing specifications and features. First question to ask yourself is, "what is my application?" Answering this question will help refine your search for a model that will be a best fit to your need.

Why People Purchase Geiger Counters:
- Safety checks for radiation levels in your environment, home, food, water, and surroundings.
- Anyone living close to a nuclear power plant.
- First responders who need a reliable and accurate Geiger counter.
- Interested in science and want to perform nuclear experiments.
- Survivalist being prepared for a nuclear accident or emergency.
- Gadget lover who wants to play with these nuclear instruments to satisfy your curiosity.
- Prospecting for uranium ore or radioactive materials.
- Collecting - checking vintage Fiesta ware plates or glow-in-the-dark clock hands for radioactivity.

What is a Geiger Counter?

Geiger counters are instruments that can detect and measure ionizing radiation using a Geiger Mueller tube. Geiger Mueller tubes are commonly referred to as GM tubes. A GM tube is the radiation sensor used in a Geiger counter.

As explained in the last chapter, radiation, as it passes through the GM tube, ionizes the gas within the tube, creating a momentary conductive path resulting in an electric pulse, heard as a 'click' sound. By measuring the number of the clicks, the instrument indicates the radiation levels.

Detection of Ionizing Radiation

Geiger counters detect the three primary radiation types; alpha, beta, and gamma (x-ray) associated with radioactivity. Radioactivity is the spontaneous emission of energy from the nucleus of certain elements, most notably, uranium.

Natural Background Nuclear Radiation:

Nuclear radiation is a normal part of our life on planet Earth. We are bombarded with nuclear radiation every day. Background radiation, from natural sources on earth and cosmic rays will cause the Geiger counter to click randomly a number of times every minute. In my corner of the world, I have a background radiation that triggers the counter 22-34 times a minute. When performing radiation checks to see if a material is radioactive or contaminated with radioactive material, this background radiation count is usually deducted from the reading to evaluate if a material is radioactive.

What Geiger Counters Do Not Detect:

Geiger counters do not detect cell phone radiation, radio frequency (RF), or electromagnetic field (EMF) radiation. EMF radiation is emitted from power transformers and other types of power electrical inductors. They cannot detect microwave radiation from a microwave oven. Nor can they detect neutrons.

Images Scientific Instruments has a free online radiation penetration calculator you can use
https://www.imagesco.com/geiger/cal/index.html

Geiger Mueller Tubes

Inexpensive GM tubes like the SBM-20 Russian Geiger counter tube shown to the right, only detect beta, x-ray, and gamma radiation. So, if your Geiger counter uses one of these styles of GM tubes it is blind to all alpha particle radiation.

Figure 2.1

Figure 2.2

More expensive GM tubes, like the LND-712 Geiger counter tube shown to the left have a thin mica window that allows alpha particle radiation to penetrate into the interior of the GM tube and ionize the gas for detection. It is generally more fragile than the beta-gamma GM tubes because of the thin mica window. The mica window allows one to detect the alpha radiation from radium and polonium. This type of Geiger counter may also be used for prospecting, experiments, and general field work. In addition, it can measure total radiation from materials including alpha radiation. Most laboratory grade Geiger counters use this style tube.

Here is another style tube that has an alpha window. This Russian tube requires 1000 VDC to operate.

Figure 2.3

The common thread connecting the types of radiation these Geiger counter tubes detect is that the radiation is all ionizing radiation. Meaning they are capable of ionizing the gas atoms inside a GM Tube which allows for their detection, as explained on the first page of this chapter.

Ebay Geiger Counters

EBAY is a place to find used and surplus Geiger counters. If you search on eBay for Geiger counters, radiation detectors, or radiation monitors, you will run across numerous radiation detection devices. Let's look at a few. There is a lot of government surplus for sale on eBay. The CD-715 model is a popular model, see Figure 2.4.

Figure 2.4

These detectors were made to be used in a high radiation field one would find in a post nuclear attack or nuclear reactor incident. The radiation meter provides the scale of detectable radiation, see Figure 2.5.

The meter scale of the CDV-715 is rated in rads/hr. A knob provided a range of 0.1 - 0.5 Rad to 100 - 500 Rads. Today's Geiger counters are far more sensitive, and measure radiation in fractions of millirads. A millirad is 1/1000 of a rad.

Figure 2.5

23

The CDV-715 is not a Geiger counter; it doesn't have a GM tube. Instead, it uses an ionization chamber. If you open a CDV-715, you can see the ionization chamber, see Figure 2.6.

Figure 2.6

The CDV-715 only detects high levels of gamma radiation, so I would not recommend this as a purchase.

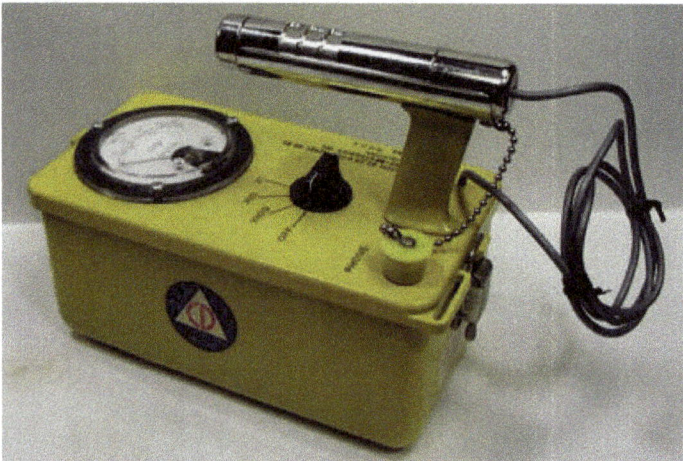

Figure 2.7

Alternatively, is the CDV-700, which looks similar, but uses a GM tube, see Figure 2.7. The CDV-700 is more sensitive than the ionization chamber CDV-715. The CDV-700 meter shows the scale, see Figure 2.8.

Figure 2.8

The GM Tube used in the CDV-700 is only beta and gamma sensitive. Since most of these units are vintage 1950-1960s, I do not have any idea as to their accuracy, or if they can be calibrated accurately. Now if you only want to purchase a working Geiger counter and can find a working unit for $50-$75.00, buy it. When you reach the $100.00 mark, you can purchase an inexpensive modern Geiger counter that has greater sensitivity.

Geiger Counter's Nuclear Radiation Readout Accuracy

Two basic types of Geiger counters are analog and digital. The digital Geiger counter is typically more accurate, as you can obtain an exact count of radioactive particles detected. Whereas, analog meter readings are averaged out, and do not provide a display for the detected particle count. A digital readout from the GCA-08 is shown in Figure 2.9 below.

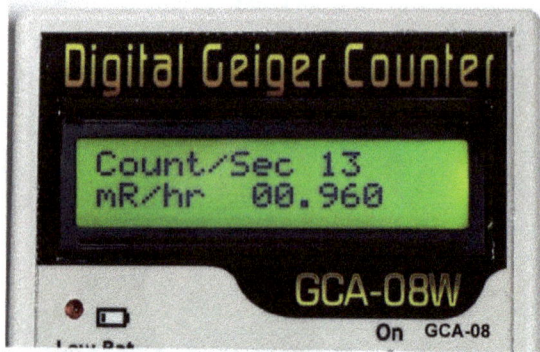

Figure 2.9

Let's discuss accuracy. Don't believe an unproven claim of accuracy. Some Geiger counter vendors will claim high accuracy for their Geiger counter, when it simply is not true. Here is what you need to know, will their Geiger counter pass ANSI-STD N323A calibration? Because without passing an ANSI-STD N323A calibration, their claim of accuracy means nothing. Just claiming that a Geiger counter is calibrated and has an accuracy of 1%, 5%, 10%, or 20% is not proof.

Fortunately, the ANSI-STD N323A calibration is the proper accuracy standard for Geiger counters. An NRC certificate, Fig. 2.10, certifies the attached Geiger counter has passed ANSI-STD N323A calibration using a NIST traceable source from an independent government licensed lab conforming to NRC regulations 10-CFR-34, 10-CFR-35, making it suitable for regulatory inspections. Certification label photo, courtesy of Images SI, Inc.

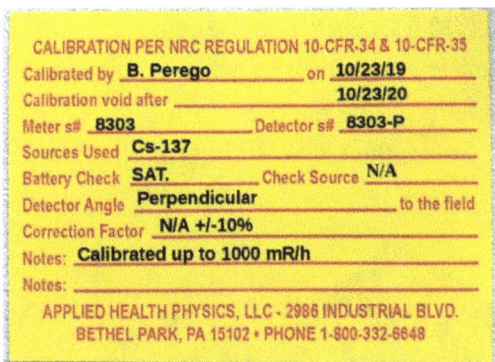

Figure 2.10
NRC certification label for the GCA-07W

Where to Obtain ANSI-STD N323A Calibration:

There are a number of independent laboratories that are licensed by the U.S. Government to calibrate Geiger counters to ANSI-STD N323A standard, and if it passes calibration, a certification label is attached to the Geiger counter. The certification is good for one year.

The laboratory my company uses is:

Applied Health Physics, LLC
2986 Industrial Blvd
Bethel Park, PA 15102
(412) 835-9555

GM Tubes: Internal or External

In the internal Geiger counter instrument, the Geiger Mueller tube is inside the case, enabling one hand operation, leaving the other hand free. Figure 2.11 shows the GCA-06 internal model.

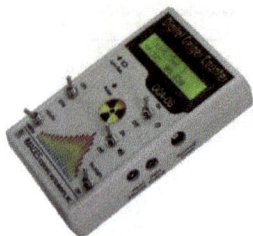 

Figure 2.11   *Figure 2.12*

For the external instrument, the tube is located outside the case at the end of a probe or wand and connected to the instrument via a cable, see the GCA-06W shown in Figure 2.12.

This kind of instrument is more suited for 'probing' radiation levels in tight spots. It's easier to move a probe around to check for radiation, than it is to move the entire instrument.

How to read a Geiger Counter

The Geiger counter measures the radiation levels in different ways. Analog meters can provide a reading in Counts per Minute (CPM) and their equivalent Radiation Level. For example, let's look at the meter face of a common analog Geiger counter, see Figure 2.13.

Figure 2.13

28

At its most sensitive scale, a count of 500 CPM is equivalent to a radiation level of 0.5 mR/hr. This model Geiger counter has a 10X and 100X switch to select ranges from 0.5 mR/hr to 5 mR/hr and 50 mR/hr respectively.

To obtain the Counts per Second (CPS), divide the CPM by 60. So, a CPM reading of 300 is equal to 5 CPS (300/60 = 5).

Digital Geiger Counters

Digital Geiger counters have a digital display. The digital display has a number of advantages, such as providing an accurate and exact count of the detected radioactive particles in either the CPS, or CPM measurement, and their equivalent radiation level. The analog display approximates the number of clicks per minute.

Figure 2.14

Figure 2.14 shows a digital display from a GCA-07, the Counts per Second (CPS) on the top line of the LCD. The second line shows the equivalent radiation level. This second line's equivalent radiation level is a running three second average. This three second average smooths out the reading so it's not jumping every second in response to the changes in the CPS.

Imperial or Metric Measurement(s)

The digital display also allows you to change from Imperial measurements (mR/hr) to Metric measurements (mSv/hr) with a flip of a switch. Figure 2.15 shows a digital display from a GCA-07, the Counts per Second (CPS) on the top line of the LCD as before. The second line shows the equivalent radiation level using metric measurements of mSv/hr. This second line's equivalent radiation level is also a running three second average, as explained before.

Figure 2.15 Switch from Imperial to Metric Measurements

CPM Measurements:

When switching to the CPM mode, the digital screen changes, see Figure 2.16.

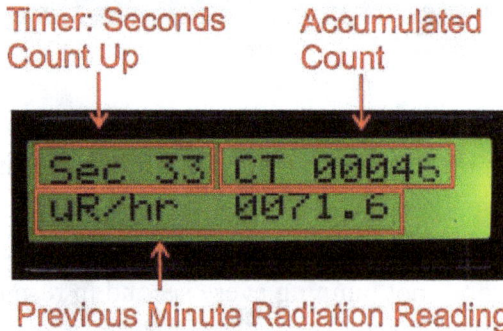

Figure 2.16

The first LCD line shows the running timer. The timer starts at 0 seconds and counts to 60 seconds (one minute). Next to the elapsed time on the first line, the accumulated count of radioactive particles detected is displayed. The timer continues to count radioactive

30

detections until a full minute (60 seconds) has passed. At this point, the display pauses for one second to show the complete count for the minute and its relative radiation level before resetting the timer and accumulated count to zero and starting to count for the next minute.

The second line will continue to show radiation level until it updates again at the end of the next minute.

One can also switch from Imperial measurement uR/hr to metric measurement uSv/hr in CPM mode.

Range:

The range of your Geiger counter tells you how much radiation you can measure. Typically, analog meter Geiger counters range from 1 to 100 mR/hr. Although some can go as high as 500 mR/hr. With digital Geiger counters, the typical range is 1 to 200 mR/hr. However, the GCA-06 and GCA-07 series of digital Geiger counters range from 1 to 1000 mR/hr and maintain their NRC accuracy.

Ease of Use:

Having a great Geiger counter that is a pain to use is no bargain. Some Geiger counters use a membrane push button to rotate through menu options to select different modes and ranges. If you read the reviews on these types of Geiger counters, most customers find this type of menu selection frustrating.

Look at Figure 2.17, all modes can be easily selected from the front panel switches. Another option that received bad reviews are membrane switches. The advantage of membrane switches is that they allow a flat profile on the Geiger counter case. The disadvantage(s) is that the microcontroller needs to see the key press when it is pressed, or the option is not initiated; and they don't have the positive feel of a standard switch closure.

You want a Geiger counter that is easy and intuitive to use. Not one that requires you to reference a manual to make mode selections.

Warranty:

The standard warranty is one year on parts and labor for any factory defects. An exception to the warranty is the GM tube, which is fragile and could be broken easily if not handled properly.

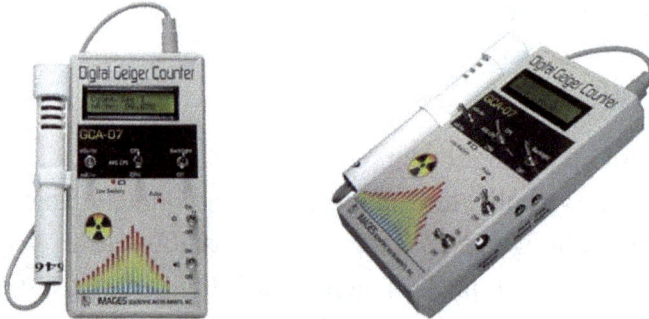

Figure 2.17

Other Options

Data Logging / Software:

Does your Geiger counter come with data logging or radiation charting software?

The radiation charting software allows one to chart radioactivity over long periods of time. It allows radiation data to be saved in files and pulled up into other programs like Excel. The radiation charting software for Images SI, Inc. is shown in Figure 2.18.

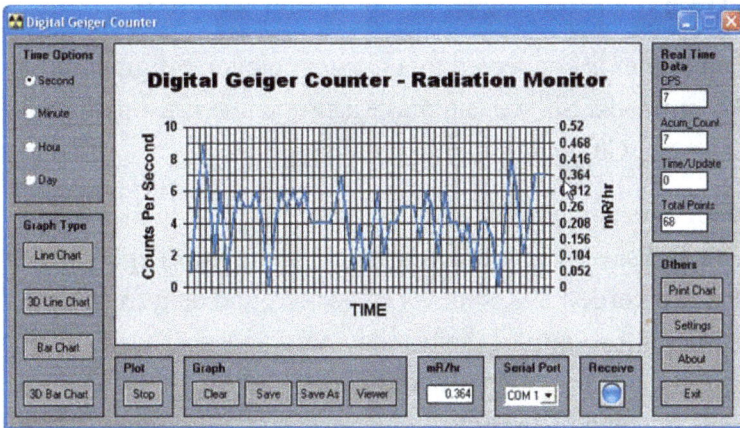

Figure 2.18

This radiation charting and monitoring software is free for downloading from Images SI Inc. website for any compatible Geiger counter. Go here: https://www.imagesco.com/geiger/geiger-graph.html

Data Loggers - Digital Output

A modern Geiger counter usually has a digital output that outputs a +3-5V pulse with each detection of a radioactive particle. A suitable Data logger that logs TTL +5 pulses is shown in Figure 2.19.

33

Figure 2.19

Connecting the Data Logger to the Geiger Counter.

This data logger is connected to a Geiger counter's digital output pulse via a 3.5 mm jack. So, we can make a simple interface using a 3.5 mm cable from the Geiger counter to data logger.

Once your data is recorded, you reconnect the data logger to the computer and save the recorded test data using the Geiger Graphing Software. I recorded 298 minutes of background data from a GCA-07. The instructions to import the test data into an Excel spreadsheet are provided in the Data Logger instructions. Once the data is imported into Excel, it can be manipulated and graphed, see Figure 2.20 below. The average Count Per Minute (CPM) was 16 with a minimum count of 6 CPM and a maximum of 28 CMP.

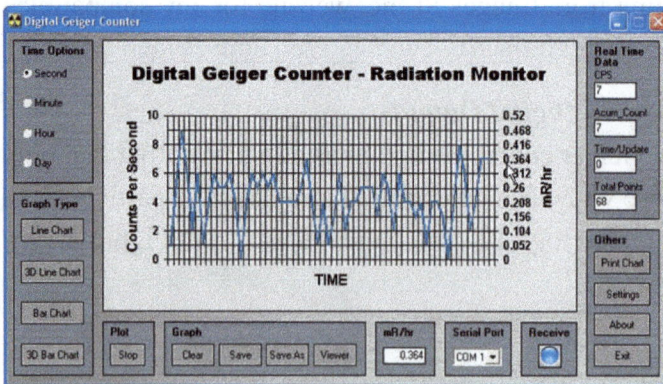

Figure 2.20

34

Plain vanilla Geiger counters that have a +5V pulse output can be brought into the Digital Age using a Digital Meter Adapter (DMAD). To work properly, the Geiger counter must output a pulse for every radioactive particle it detects.

Figure 2.21

The DMAD Adapter (Chapter 16) may be used with any Geiger counter that outputs a TTL Digital pulse for each radioactive particle detected. Since the DMAD is a general device, the radiation conversion may not be exact since different GM tubes have different responses to radiation. The CPS or CPM counts will be exact. The DMAD also has a true random number generator function.

Figure 2.22

The DMAD may be purchased as a kit or assembled and tested. Figure 2.23 shows the DMAD connected to an inexpensive analog Geiger counter.

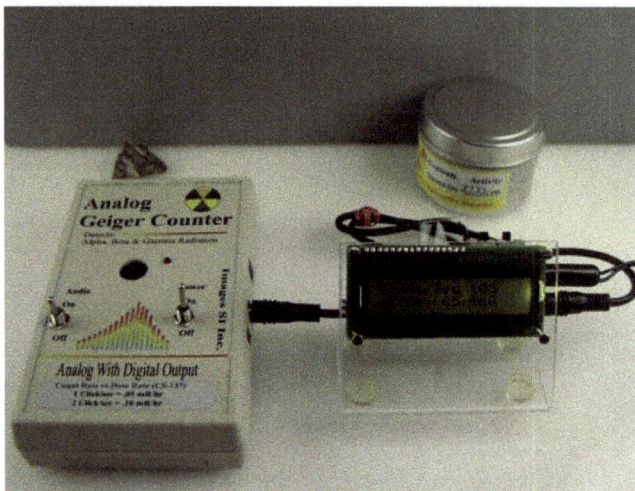

Figure 2.23

The DMAD also has a serial output that is compatible with the USB-TTL cable that will allow you to use the Images SI, Inc. free Radiation Monitoring and Charting Software.

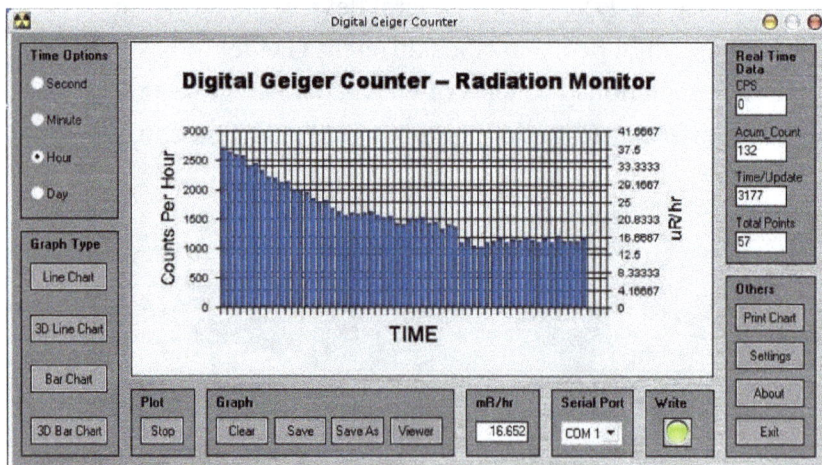

Figure 2.24

Let's review the options covered in this chapter.

- Determine your interest and application.
- What radiation do you need to detect?
- Geiger counter accuracy. Can it be NRC Certified?
- Digital or Analog - Imperial/Metric or Both
- Range
- Data Logging and Radiation Charting Software
- Ease of Use
- Working within your budget
- Warranty

I could not obtain permission from some manufacturers to list their Geiger counters. Therefore, I was limited in the Geiger counters I could feature.

Chapter 3

3 - Introduction to the GCA-07W Geiger Counter

The experiments performed in this book used a digital GCA-07W Geiger counter that passed ANSI-STD N323A calibration. Any digital Geiger counter with similar features may be used to conduct the experiments.

Scientific Instrument & Industrial Tool

The free Windows graphing program is available from the company website. Software requires a compatible USB to 3.5mm TTL cable. Numerical information from graph files may be exported to Excel spreadsheets.

Communication specifications are provided for users to read the output of the Geiger counter and write their own programs.

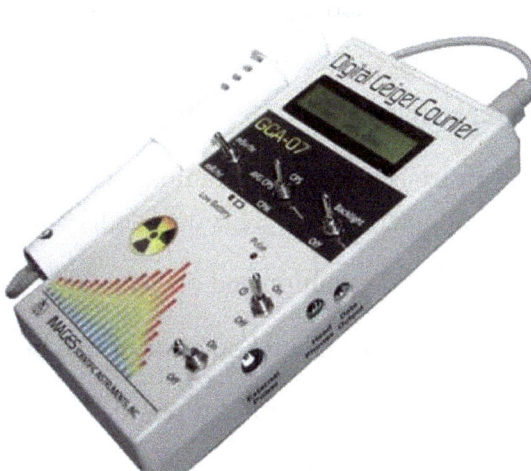

Figure 3.1

41

Features

Radiation Detected: Alpha, Beta, Gamma, and X-Rays.

Detector: Geiger-Mueller tube Ne + Halogen filled with a .38" effective diameter 1.5-2.0 mg/cm2 mica end window.

Detector Sensitivity

- Alpha above 3.0 MeV
- Beta above 50 KeV
- Gamma above 7 KeV
- Countable Pulse Range:
  1 (CPM) – 10,000 + counts per second (CPS)
- Converted Radiation Range:
  .05 mR/hr - 1000 mR/hr (Imperial)
  .0005 mSv/hr - 10.0 mSv/hr (SI Metric)

The **Liquid Crystal Display (LCD)** is a 16 character by 2 line that provides an easy to read output, see image to right. LCD has an on-off backlight switch.

Backlight Switch turns on and off the LCD backlight.

The LED marked **Low Battery** will turn on when the battery power drops and needs replacing.

The LED marked **Pulse** is a secondary radioactive particle indicator. It blinks each time a radioactive particle is detected.

The **Power Switch** turns the unit on and off.

The **Speaker Switch** turns the sound on or off to the internal speaker.

Headphone jack is a standard 3.5mm for private listening.

External power jack is available for extended readings. Transformer is included.

TTL Serial output for PC available via 3.5MM stereo connector.

Main Panel Controls

The first panel switch starting from the left selects whether the radiation levels are shown in Systems International (SI) metric (mSv/hr) or imperial (mR/hr) measurements.

Figure 3.2

The middle switch labeled CPS, AVG CPS, and CPM selects one of the three Survey Meter Modes:

Figure 3.3

CPS (Counts per Second) is a one-second counting mode. Real time radiation readings display the count/second and equivalent radiation level in either mR/hr or mSv/hr.

43

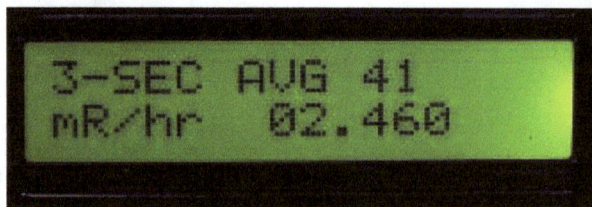

Figure 3.4

AVG CPS is a three-second average of the CPS. AVG CPS performs a smoothing function similar to analog meter readings. It displays the 3 Second CPS and equivalent radiation level in either mR/hr or mSv/hr.

CPM (Counts per Minute) is a one-minute counting mode for measuring low levels of radioactivity and background radiation: Displays accumulated count and equivalent background radiation in either uR/hr or uSv/hr. If radiation level is significant, radiation scale level changes and is displayed either mR/hr or mSv/hr.

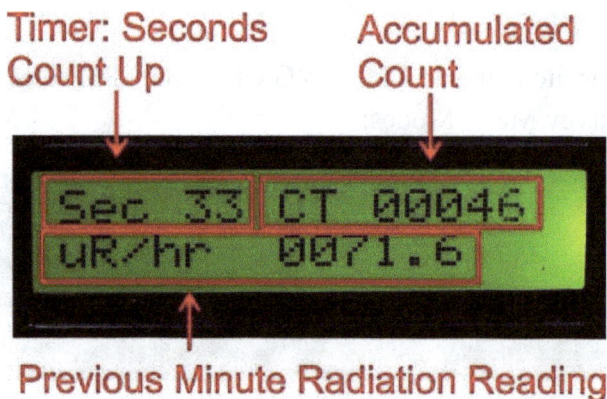

Figure 3.5

The first line of the LCD display shows the accumulated time in seconds. The timer starts at 0 seconds and counts to 60 seconds (one minute). When it reaches 60 seconds, it pauses one second then resets. Next to the elapsed time is the accumulated count of radioactive particles detected. The timer continues to count radioactive detections until a full minute (60 seconds) has passed. At this point the display pauses for one second to show the complete count for the minute and

44

its relative radiation level before resetting the timer and accumulated count to zero and starting to count radioactive particle detections for the next minute, CPM measurement.

The second line on the LCD will continue to show radiation level until it updates again at the end of the minute.

One can also switch from Imperial measurement uR/hr to metric measurement uSv/hr in CPM mode. Updates in imperial/metric mode initiate at the end of the current CPM measurement, when the timer and accumulated count resets.

Operations

Survey Meter Modes

CPS Mode: Set the Conversion switch to mR/hr (milliroentgen/hour). The time function switch to "CPS", Backlight switch ON, and the audio switch ON. Turn on the Geiger counter. If you have a radioactive source, bring the source close to the GM tube. For Geiger counters with an external wand, bring the wand close to the radioactive source.

Figure 3.6

Every radioactive particle detected will cause the Geiger counter to click and the LED to blink. The LCD digital display in this mode updates the count and radiation level every second, see photo above. The display always shows the previous seconds count and radiation level. The count "Count/Sec" is the number of radioactive particles detected in the previous second. On the second line is the equivalent

45

radiation level of that count in mR/hr. You can change the Conversion switch to mSv/hr to read the radiation level in milli-sieverts/hour.

3-Second Average: The three-second average of the CPS. AVG CPS performs a smoothing function similar to analog meter readings. Displays the 3-Second CPS and equivalent radiation level in either mR/hr or mSv/hr.

3-Second Average of
Count Per Second

3-Second Average
Current Count

3-SEC AVG 41
mR/hr 02.460

3-Second Average
Radiation Reading (Imperial)

Figure 3.7

CPM mode: Displays the counts per minute and converts the radiation level into micro-Roentgens (uR/hr) or micro-Sieverts (uSv/hr). The CPM mode is useful for checking background radiation. First, set the switch to Metric or Imperial measurement. Next, set the time function switch to CPM. The LCD display changes. The left side of the first line begins a count starting at 1 and increments by 1 up to 60 seconds.

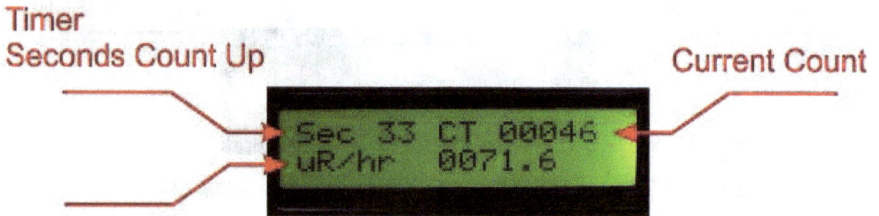

Timer
Seconds Count Up

Current Count

Sec 33 CT 00046
uR/hr 0071.6

Previous Minute
Radiation Reading

Figure 3.8

Free Windows PC Geiger Graphing Software

So far, you needed to record CPM numbers manually. However, there is an easier way. Images SI, Inc. offers a free download of their Windows PC Geiger Graphing software (https://imagesco.com/geiger/geiger-graph.html). This software will chart and record an unlimited number of readings. So, you could take an hour or more CPM background readings and save the readings to your hard drive.

To connect the digital Geiger counter to the PC requires a **compatible** 3.5mm TTL to USB cable.

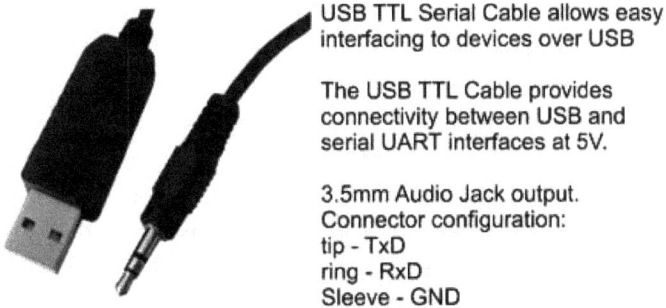

USB TTL Serial Cable allows easy interfacing to devices over USB

The USB TTL Cable provides connectivity between USB and serial UART interfaces at 5V.

3.5mm Audio Jack output.
Connector configuration:
tip - TxD
ring - RxD
Sleeve - GND

Figure 3.9

This software allows you to graph radioactivity in a number of different time scales. The data collected on the graph can be saved and converted to a CSV format that can then be imported in Excel for further analysis and graphing.

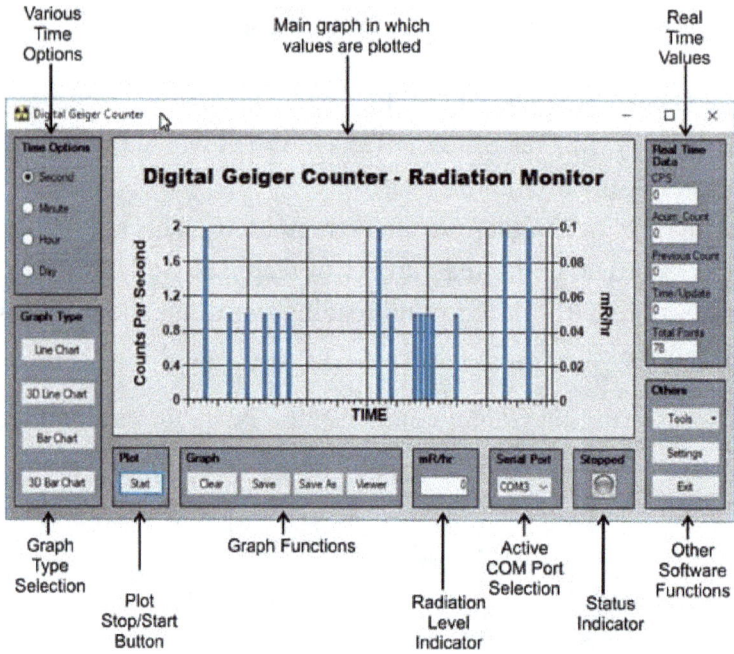

Figure 3.10

The time scale shown below is in CPM and shows the background radiation for sixty CPM readings.

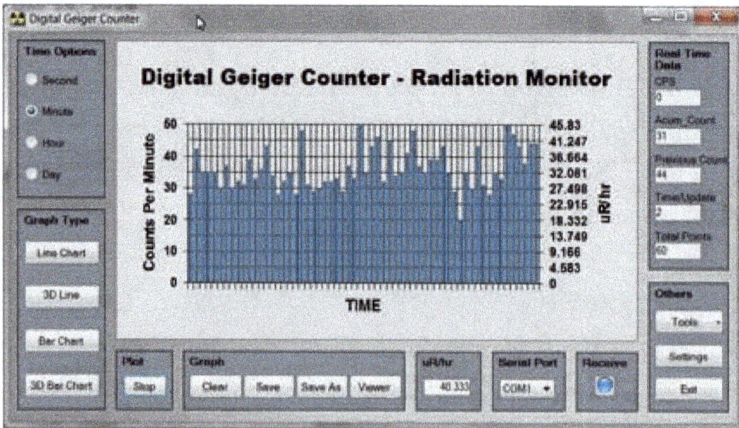

Figure 3.11

Chapter 4

4 - Measuring Background Radiation

Sources of Background Radiation

Everyone on the Earth is exposed to background radiation. Therefore, it is important to establish the sources of our background radiation. The three primary sources are radiation from space (cosmic rays), terrestrial (Earth), and internal (for our own body). Cosmic radiation from deep space, and some released from our Sun during solar flares, account for 8% of natural radiation exposure. Terrestrial radiation from rocks and soil accounts for 8% of natural radiation exposure. Added to the terrestrial radiation is radon. Radon, which is an invisible heavier than air gas, emitted by uranium and thorium, accounts for about 65% of natural radiation exposure. Finally, internal radiation; our bodies also contain radioactive materials, like potassium-40, which account for approximately 11% of natural radiation exposure.

Why Measure Background Radiation?

In order to make accurate measurements for some experiments, you may need to subtract the background radiation from your measurements in order to obtain meaningful results.

For example, if you are measuring material that is emitting low levels of radioactivity, you would need to subtract the background radiation to determine the radioactivity level of the material.

Going a little further, suppose you needed to check to see if a material is radioactive at all. Here not only do you need to know the background radiation, but the standard deviation which would help to determine if the radioactivity level you are reading is high or a normal fluctuation in the background radiation.

To take a measurement, we first need to read our Geiger counter instrument.

How to set up and read a Digital Geiger Counter

The GCA-07, GCA-06, and GCA-03 Series of digital Geiger counters have similar digital displays. The digital display has a number of advantages over analog meters, such as providing an exact count of the detected radioactive particles and their equivalent radiation level.

While the picture shows the front panel of the GCA-07W, the other series of Digital Geiger counters have similar controls.

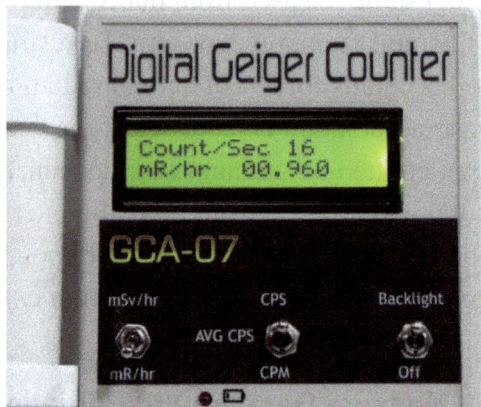

Figure 4.1

For more information on the use and function of the GCA-07W, refer to the previous chapter.

Taking Background Radiation Measurements

We use the CPM mode of the Geiger counter to take background radiation measurements. Position the Geiger counter or the Geiger counter's sensor in an area that is free from external radioactive sources. Take a minimum of 10 CPM measurements and log the CPM number in a notebook. In the example shown below, 20 CPM measurements were recorded then added together for a total. Divide the total by the number of samples, in this case 20. This will be your average CPM or background radiation. In addition, note the highest and lowest CPM numbers you obtain in your sample. This will be your Max and Min.

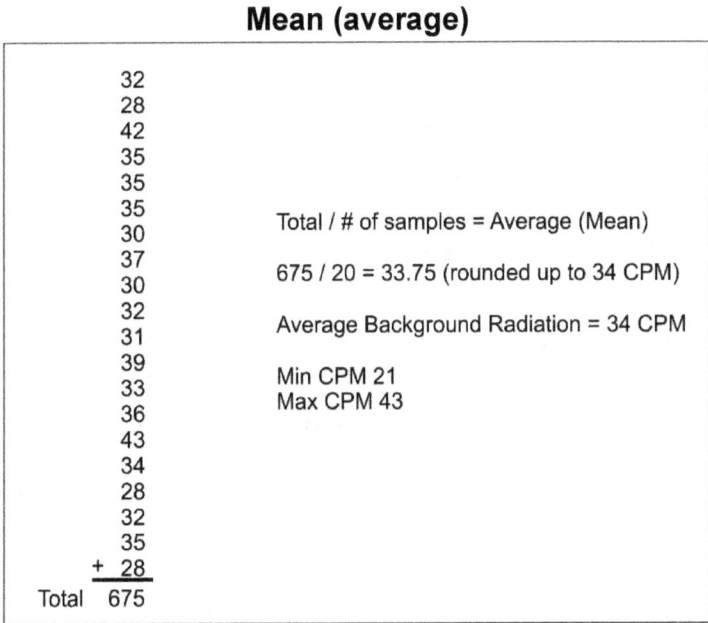

Mean (average)

```
        32
        28
        42
        35
        35
        35             Total / # of samples = Average (Mean)
        30
        37             675 / 20 = 33.75 (rounded up to 34 CPM)
        30
        32             Average Background Radiation = 34 CPM
        31
        39             Min CPM 21
        33             Max CPM 43
        36
        43
        34
        28
        32
        35
      +  28
Total   675
```

Figure 4.2

In general, the higher the number of samples you record, the greater the accuracy.

You can use this basic background radiation number to check objects in your environment for radioactivity. Place the wand of the Geiger counter on or near the suspect object and take a few CPM readings. For instance, I placed the wand on the screen of a working old-style B/W television. This type of television uses a cathode ray tube to create a picture. My CPM reading was 65. Using the information above, you could conclude that the Cathode ray tube may be emitting soft x-rays.

The time scale shown above is in CPM and shows the background radiation for sixty CPM readings.

To finish this section, take 10 or 20 CPM readings from your Geiger counter. Keep the Geiger counter or sensor away from any external source of radioactivity. Total the CPM readings and divide by the number of samples to obtain your average background CPM.

Part 2. Statistical Analysis of Background Radiation

In addition to calculating the average background radiation is calculating the sample standard deviation. The standard deviation quantifies the variance in the CPM readings. The standard deviation determines how far a CPM reading is from the mean (average). This can be useful when performing experiments where one may be trying to determine where a particular CPM reading falls, whether it is reasonably inside the sample group or outside the sample group.

The chart below illustrates the general standard deviation. From the graph, 68% of values are found within 1 standard deviation from the mean. 95% of values are found within 2 standard deviations from the mean. Finally, 99.7% of values are found within 3 standard deviations from the mean.

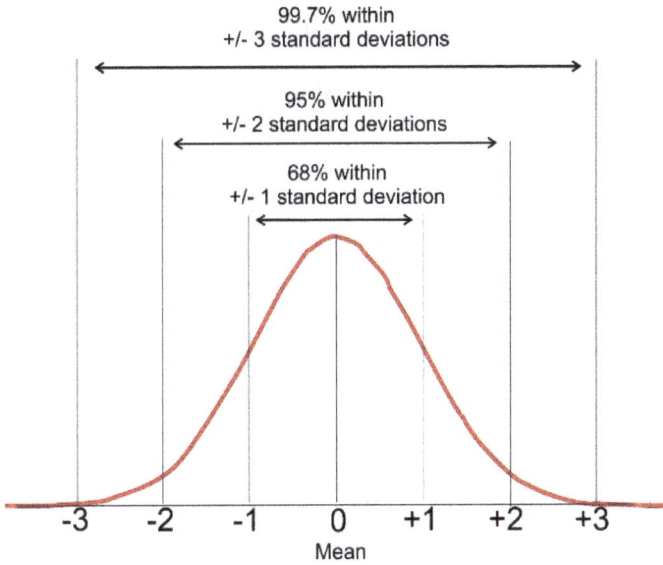

Plot Standard Deviation
Figure 4.3

The formula for the sample standard deviation follows. The symbol s is the sample standard deviation. The summation sign iterates each CPM value 1 through 20 through the formula and sums them together. The variable x_1 are the CPM values. The symbol $\bar{x}$ is the sample mean (average). The results of (x_1 - $\bar{x}$) are squared (x_1 - $\bar{x}$)^2 then added to sum. When all CPM numbers have gone through the formula, and are summed together, that result is multiplied by 1/20-1 or 1/19. Taking the square root of that result is the standard deviation.

Sample Standard Deviation

$$S = \sqrt{\frac{1}{N-1}\sum_{i=1}^{N}(x_1 - \bar{x})^2}$$

How to Calculate Mean and Standard Deviation Manually:

Step 1) Determine mean of the CPM test results.

Step 2) Calculate difference between mean and individual test results.

55

Step 3) Calculate the square of individual results from Step 2.

Step 4) Determine the sum of the values that were calculated in Step 3.

Step 5) Calculate the standard deviation by dividing the sum from Step 4 by the number of tests minus 1, and then get the square root of that number.

Step 1 (list)	Step 2 (CPM(x) - Mean)	Step 3 (CPM(x) - Mean)^2
(x = CPM count)	(x - 33.75)	(x - 33.75)^2
32	-1.75	3.06
28	-5.75	33.06
42	8.25	68.06
35	1.25	1.56
35	1.25	1.56
35	1.25	1.56
30	-3.75	14.06
37	3.25	10.56
30	-3.75	14.06
32	-1.75	3.06
31	-2.75	7.56
39	5.25	27.56
33	-0.75	.56
36	2.25	5.06
43	9.25	85.56
34	.25	.06
28	-5.75	33.06
32	-1.75	3.06
35	1.25	1.56
+28	-5.75	+33.06
Total = 675		Sum 346.14

(Average) Mean = 675 / 20 = 33.75

(Step 1) To get the average, divide the Total by the number of samples. (Step 2) Then subtract the calculated average from each sample. (Step 3) Square the results of Step 2. (Step 4) Sum all results in step 3. (Step 5) Use the sum result to calculate the sample standard deviation using the following equation:

Step 5 $\sqrt{\frac{1}{19} * 346.14}$, $\cong \sqrt{18.21}$ = 4.27. The standard deviation for these 20 CPM samples is 4.27. So, for this group, the mean is 33.75, and the standard deviation is 4.27

Using Excel to calculate Mean and Standard Deviation

Obtain 10 or 20 sample CPM readings from your Geiger counter. If you are using Images Geiger Graphing software, you can save CPM data to a file that can be directly imported into Excel. If not, you can enter the data manually.

Put the CPM values in column A. In the screenshot below, column A1 to A20 is filled with the 20 CPM values from our sample. Choose a cell to display the Standard Deviation result. In the figure below, I chose cell D6. In the formula field above, see red box, enter the following formula =STDEV.S(A1:A20) and hit return. The cell will display the standard deviation. The Standard Deviation matches our manual calculation.

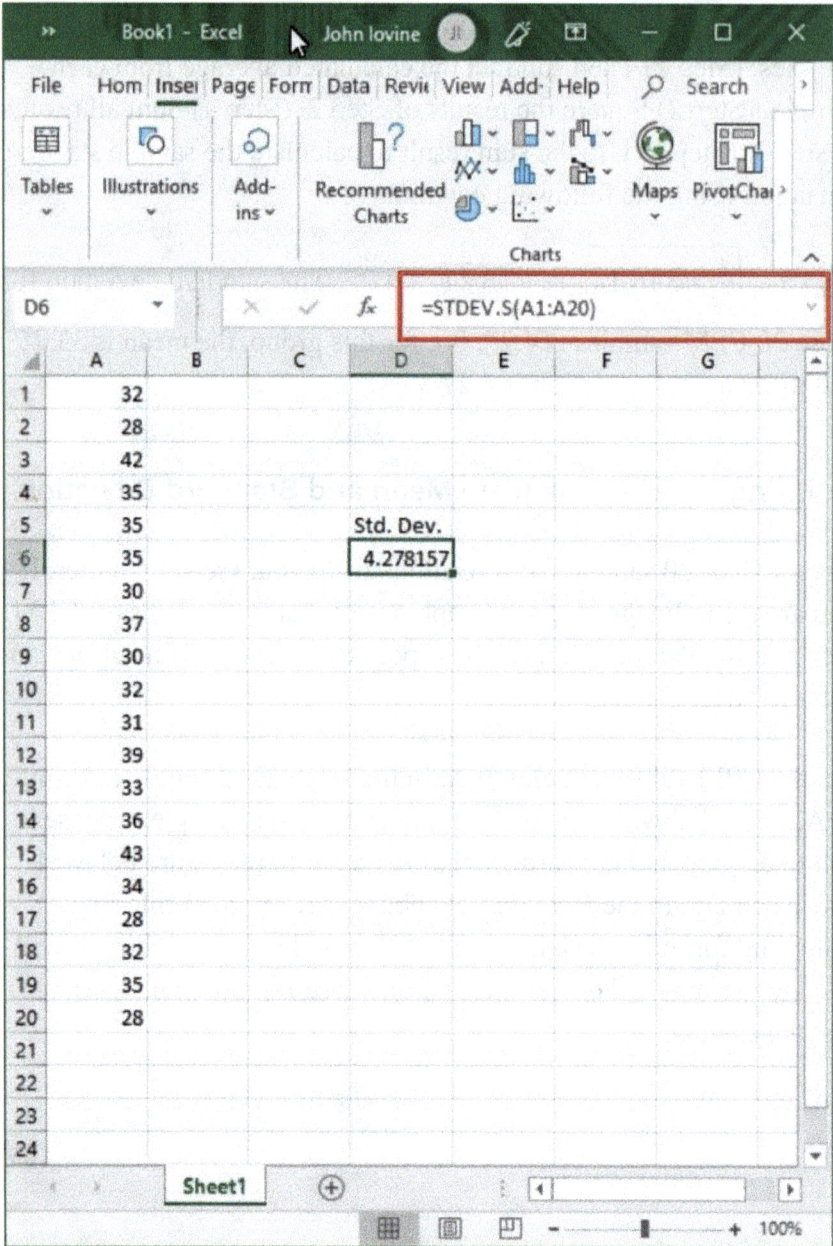

Figure 4.4

Going Further with Excel

You can do more with Excel than what is shown. For instance, Excel is also capable of plotting the standard deviation of the data points collected.

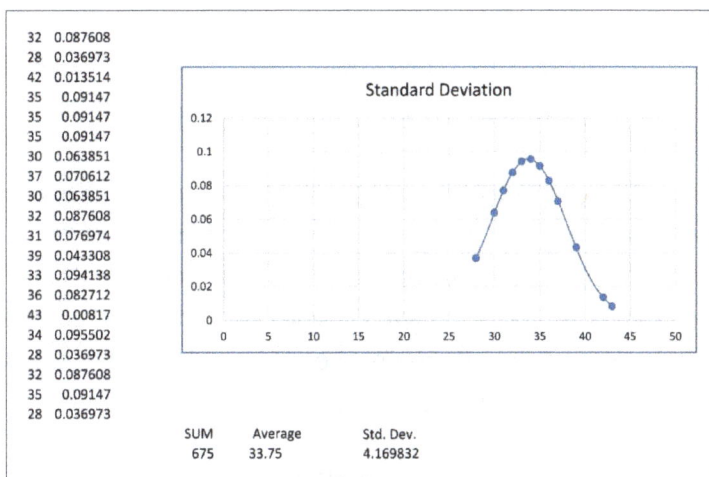

32	0.087608
28	0.036973
42	0.013514
35	0.09147
35	0.09147
35	0.09147
30	0.063851
37	0.070612
30	0.063851
32	0.087608
31	0.076974
39	0.043308
33	0.094138
36	0.082712
43	0.00817
34	0.095502
28	0.036973
32	0.087608
35	0.09147
28	0.036973

SUM	Average	Std. Dev.
675	33.75	4.169832

Standard Deviation

Figure 4.5

Z-Score

The z-score is the calculation of how many standard deviations a number is from the mean. For example, suppose we obtained a reading of 27 CPM. The value of 27 CPM would have a z-score of 1.61, since the value of 27 is 1.61 standard deviations from the mean.

To convert a value to its z-score, do the following:

Subtract the value from the mean $33.75 - 27 = 6.75$

Divide that result by the standard deviation $6.75 / 4.17 = 1.61$

The z-score formula

$$z = \frac{x - \bar{x}}{s}$$

z is the z-score
x is the value
x̄ is the mean
s is the standard deviation

Practical Applications - Example 1

In the case where we were checking an unknown material to see if it was radioactive: Assume we were reading CPM counts in the 50 to 60 CPM range. We know our mean is 33.75 CPM, rounded off to 34 CPM. Our standard deviation is 4.27. The counts we are reading are more than 3 standard deviations from our mean. We could say with a high degree of confidence that the material is radioactive.

Practical Applications - Example 2

In the case where we were checking a second unknown material to see if it were radioactive: Assume we were reading CPM counts in the 40 CPM range. We know our mean is 33.75 CPM, rounded off to 34 CPM. Our standard deviation is 4.27. The counts we are reading are within 2 standard deviations from our mean. We could reasonably conclude that the 40 CPM reading could be a normal fluctuation in our background radiation.

Practical Application - Example 3

In the case where I checked for radiation being emitted from an old cathode ray TV tube and obtained a CPM reading of 65 CPM: This reading falls outside 3 standard deviations from the mean. We could reasonably conclude that the cathode ray tube is emitting soft x-rays.

Online Calculators for Performing Math

Convert Microcurie to Becquerel

https://www.unitconverters.net/radiation-activity/microcurie-to-becquerel.htm

Exponents

https://www.calculatorsoup.com/calculators/algebra/exponent.php

5th Roots calculator

https://www.calculatorsoup.com/calculators/algebra/fifthroots.php

Euler's Number **2.718281**

Chapter 5

5 - Poisson Distribution

The Poisson distribution equation will calculate the probability of a given number of events happening within a fixed amount of time.

To see how the Poisson distribution works, we can use our data from the last chapter. Our CPM average was 33.75. The mean value of 33.75 may be identified as either μ or λ. To get our Counts per Second (CPS) rate, we divide our CPM number by 60. This works out to 0.5625 counts per second. Using Poisson, we can calculate the probability of reading 2 counts in a second (n = 2).

$$P(n) = e^{-\mu}\frac{\mu^n}{n!}$$

The probability of **n** occurrences, where n = 2 and where **u** (the mean value of events) is 0.56. And **n!** is n factorial.

Our first term

$$e^{-\mu}$$

The term e is Euler's Number **2.718281** and **u** is the mean. Plugging in the numbers:

$$2.7182^{-0.56} \quad \textbf{works out to} \quad \frac{1}{2.7182^{0.56}} = \textbf{0.5712}$$

Our next term

$$\frac{\mu^n}{n!} \text{ works out to } \frac{.56^2}{2!} = \frac{.3136}{2} = \mathbf{0.1568}$$

Plugging these numbers in we get 0.5712 * 0.1568 = 0.089 or about 9% probability.

Using the formula above and the data from the previous chapter determine the probability of CPM greater than or equal to a count of 40.

How do we work out this equation? Well, we can figure out the probabilities of the number 40 and all numbers above 40.

$$P(n > 39) = P(n = 40) + P(n = 41) + P(n = 42)...\infty$$

Since there is an infinite amount of numbers above 40, this is not a practical approach.

We know all probabilities equal 1. So, if we calculated the probabilities of all the numbers under 40, 0 to 39, then subtracted those probabilities from 1, we'd have our answer.

$$1 - \left[P(n \le 39) = P(n = 0) + P(n = 1) + P(n = 2)...P(n = 39) \right]$$

If we calculate all the values to 3 decimal places, we obtain a table like this:

N   Probability
0   0.000
1   0.000
2   0.000
3   0.000
4   0.000
5   0.000

6	0.000
7	0.000
8	0.000
9	0.000
10	0.000
11	0.000
12	0.000
13	0.000
14	0.000
15	0.000
16	0.000
17	0.001
18	0.001
19	0.002
20	0.003
21	0.005
22	0.008
23	0.012
24	0.017
25	0.023
26	0.030
27	0.037
28	0.045
29	0.052
30	0.058
31	0.064
32	0.067
33	0.069
34	0.068
35	0.066
36	0.062
37	0.056
38	0.050
39	0.043

0.84

Total Probability 0.84 for obtaining a count of 39 and under.

Subtract this result from one, 1- 0.84 = 0.16 % probability of obtaining a CPM count of 40 or above.

Does this probability of 16% match the 40 or above CPM occurrence in the raw data?

Last question, what is the probability of obtaining a CPM count of 42?

Here's how the equation lays out:

$$P(n) = e^{-\mu} \frac{\mu^n}{n!}$$

$$P(42) = e^{-33.75} \frac{33.75^{42}}{42!}$$

If you do the math, the probability works out to 2.4%

Chapter 6

6 - Shielding - Alpha, Beta, and Gamma Radiation

Nuclear radiation is ionizing radiation. Ionizing radiation is radiation that can strip electrons from atoms and molecules. We classify this ionizing radiation into three major categories: gamma rays, beta particles, and alpha particles.

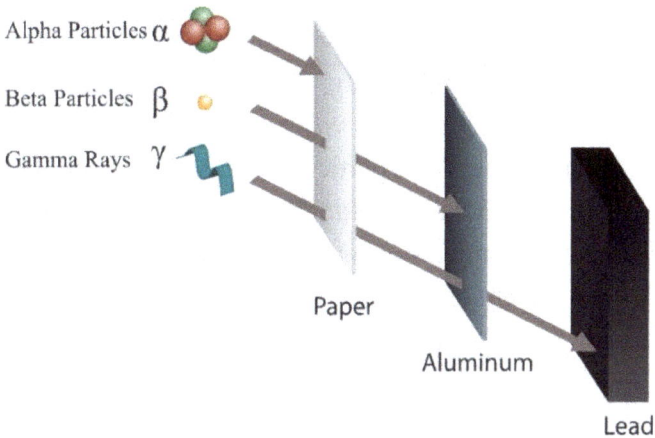

Figure 6.1

Gamma (and x-rays) are ultrashort electromagnetic radiation. They have great penetrating power and can easily pass through the body. They are attenuated by dense materials such as lead.

Beta particles are electrons. Beta particles have a net negative charge, and they have low penetrating power. Most beta radiation can be blocked by 1/8" (4mm) of aluminum.

Alpha particles are massive particles consisting of two neutrons and two protons. Alpha particles have a net positive charge. They are

71

equivalent to the nucleus of a helium atom. Alpha particles have a low penetration power. A few inches of air or a piece of paper can effectively block alpha particles.

In this experiment, we will examine each type of radiation using a Geiger counter and verify its penetrating power using paper, aluminum, and lead shields. License exempt radioactive isotope button sources are available for purchase. You will need alpha, beta, and gamma button sources.

To begin, take five or more background CPM measurements and record the results.

Background <u>19</u> CPM

The first source we will use is the polonium-210 alpha particle source. The polonium-210 has a half-life of 138 days. After the source has been in storage for a year or so, it will become too weak to perform useful experiments. The alpha particle source will need to face the mica window of the GM tube inside the wand. It will need to be in close proximity to the mica window. I needed to move my source to a distance of only 1.5 cm away from the mica window of the GM tube to get a significant CPM reading above background.

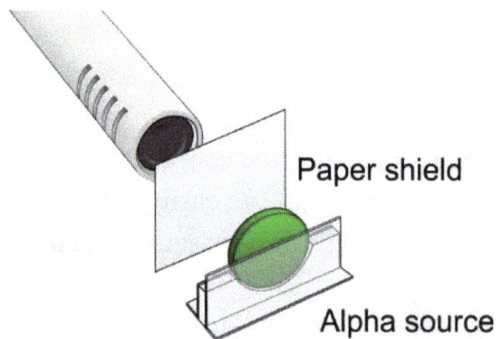

Figure 6.2

With no shield, the average CPM from the alpha source is 68 CPM.

Once you obtain a reading on the Geiger counter, place a piece of paper in-between the source and GM tube. Record your results.

Next, use the Sr-90 beta source. I placed this source 4 cm from the front of the GM tube.

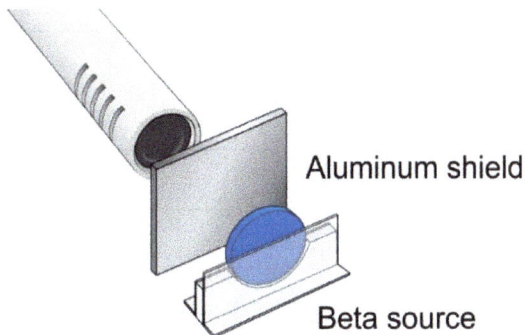

Figure 6.3

Take the readings as before and record your average CPM in chart. Next, place the paper shield in front of the beta source. Record your results. Next, place an aluminum shield of approximately 1/8" (4mm) thickness in-between the source and GM tube. Record your results.

Next, use the Co-60 gamma source. I placed this source 3 cm from the front of the GM tube. Take CPM readings as before. Record your results. Next, place an aluminum shield of approximately 1/8" (4mm) thickness in-between the source and GM tube. Record your results. Next, place a lead shield of approximately 1/8" (4mm) thickness in-between the source and GM tube. Record your results. Next, place a lead shield of approximately 1/2" (13mm) thickness in-between the source and GM tube. Record your results.

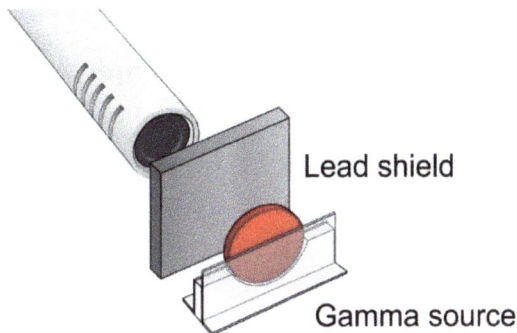

Figure 6.4

Record Your Results:

Background:	CPM		
Alpha Source	Type:	Distance:	
	No Shield		CPM
	Paper Shield		CPM
Beta Source	Type:	Distance:	
	No Shield		CPM
	Aluminum Shield		CPM
Gamma Source	Type:	Distance:	
	No Shield		CPM
	Aluminum Shield		CPM
	Lead Shield 0.125		CPM
	Lead Shield 0.70		CPM

What conclusions can you draw about the penetrating power of radiation?

If you were supplied with a radioactive sample, could you determine the percentages of alpha, beta, and gamma radiation being emitted from the source?

Test Results

Background: 19 CPM		
Alpha Source	Type: PO-210	Distance: 1.5cm
	No Shield	68 CPM
	Paper Shield	19 CPM
Beta Source	Type: SR-90	Distance: 4cm
	No Shield	366 CPM
	Aluminum Shield	19 CPM
Gamma Source	Type: CO-60	Distance: 3cm
	No Shield	213 CPM
	Aluminum Shield	178 CPM
	Lead Shield 0.125	146 CPM
	Lead Shield 0.70	90 CPM

Chapter 7

7 - Inverse Square Law Experiment

In this experiment, we will examine the Inverse Square Law. The law states that the intensity of radiation is inversely proportional to the square of the distance from the source.

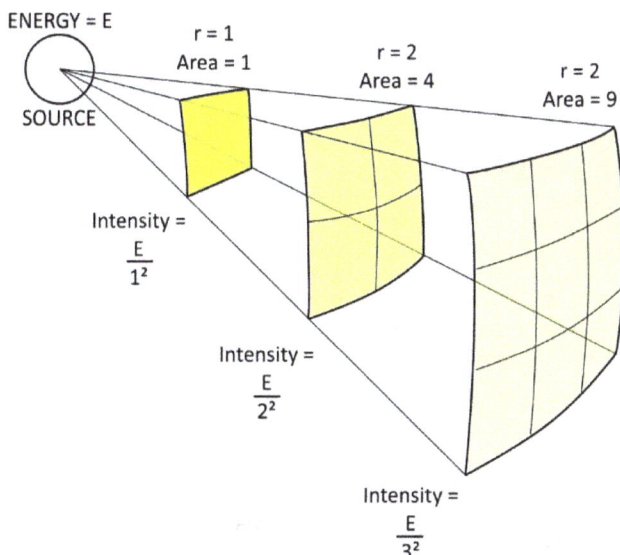

ENERGY = E

r = 1
Area = 1

r = 2
Area = 4

r = 2
Area = 9

SOURCE

Intensity =
$\dfrac{E}{1^2}$

Intensity =
$\dfrac{E}{2^2}$

Intensity =
$\dfrac{E}{3^2}$

Figure 7.1

I used a CS-137 10 uCi source for this experiment. I started measuring CPM counts at 16 centimeters from the source using the side of the GM tube. To obtain accuracy for this experiment, the distance between the source and the detector (GM tube) needs to be several times greater than the diameter of the GM tube. The diameter of the GM tube is approximately 1.5 cm (.59").

Performing the experiment:

Mark the distances away from the side of the Geiger counter wand. Place the gamma ray source at each distance mark and take five CPM readings. Average the results and fill in the table below. Once the table is filled, perform the calculations; d^2, $1/d^2$. Use this information to graph your results. Once the results are graphed, it should resemble the graph provided below. Notice the red line doesn't go through all the plotted points. Instead it is a straight line that best fits to the plotted points.

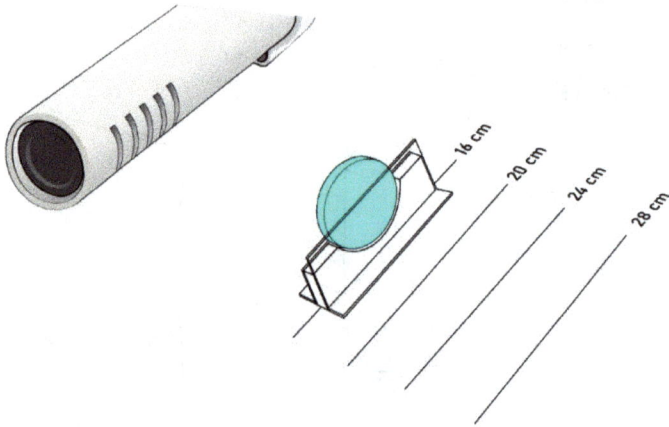

Figure 7.2

Distance d (cm)	d^2	$1/d^2$	CPM

Below are the results I obtained. Notice that if you extend the red line, it will not cross the y-axis at zero. Can you guess a reason why this is so?

If you took background CPM readings and subtracted them from every distance reading, do you think the red line would cross closer to zero?

Extending this logic further, in the chart below, the point where the red line would cross the Y-axis represents what value?

Distance d (cm)	d²	1/d²	CPM
16	256	39 x 10⁻⁴	238
20	400	25 x 10⁻⁴	147
24	576	17 x 10⁻⁴	112
28	784	13 x 10⁻⁴	79
32	1024	10 x 10⁻⁴	71
36	1296	8 x 10⁻⁴	61

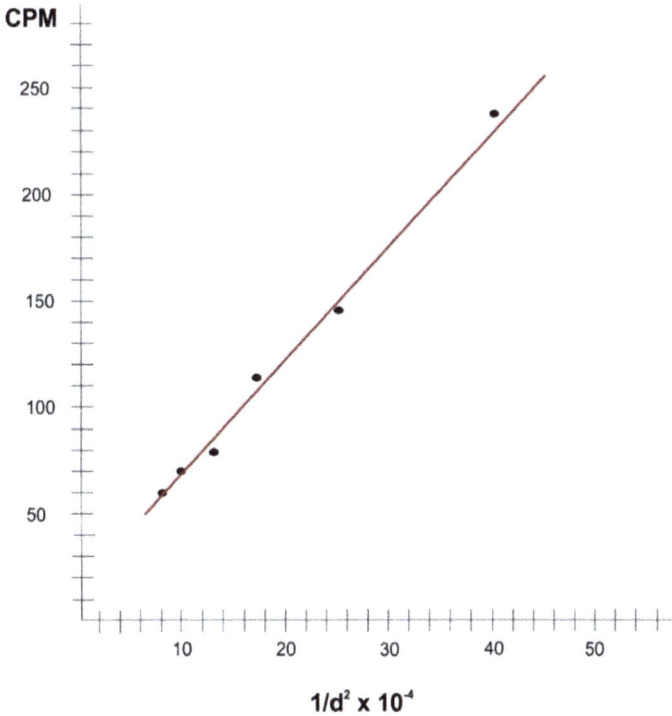

$1/d^2$ x 10^{-4}

Figure 7.3

Chapter 8

8 - Detecting Low Level Radioactivity in Food

Demonstration using common Potassium Chloride (KCl) Salt substitute

Figure 8.1

Typically, a scintillation counter is used to check food(s) for radioactive contamination. The scintillation counter not only provides a radiation level, it also provides a gamma ray spectroscopy for radioisotope identification and therefore can determine the safety of the food.

While our Geiger counter cannot perform radioactive identification, it may be used to test if a material or food stuff has a radioactivity above normal background radiation. This is not a safety check for food. Why? Because some radioactive contamination may be hidden in the normal fluctuations of background radiation. And as we will see, while KCl has a radioactivity level well above background radiation, it is commonly sold over the counter in supermarkets across the country as a salt substitute for people who are sodium sensitive.

Performing the Experiment:

Step 1) Begin by taking a set of background radiation readings. Set GCA-07 or GCA-07W to CPM mode using the front panel switch. Log a minimum of ten one-minute readings. If you are using the radiation graphing program included with the CGA-07(W), start the program. Set the time increment to one minute. The chart below shows background readings for 60 minutes.

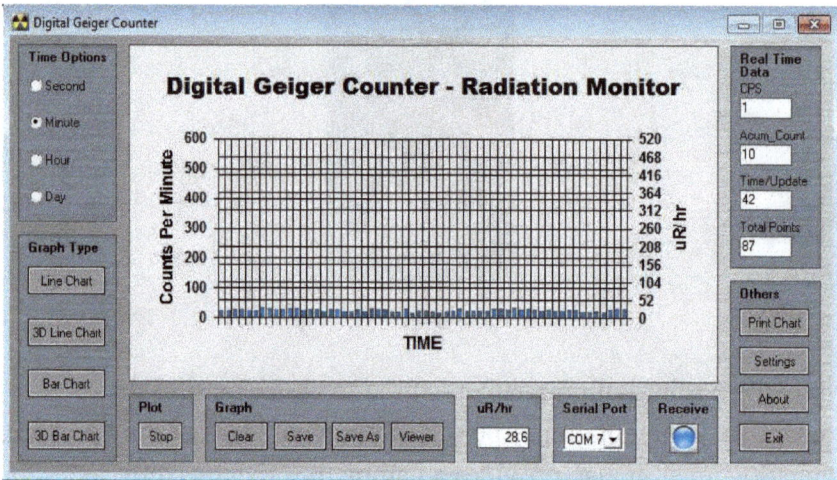

Figure 8.2

The readings may be a bit difficult to see in the standard chart program. The data may be saved, and then uploaded into the viewer program.

Images Radiation Graphing program allows you to save data and reload it into its own data viewer program.

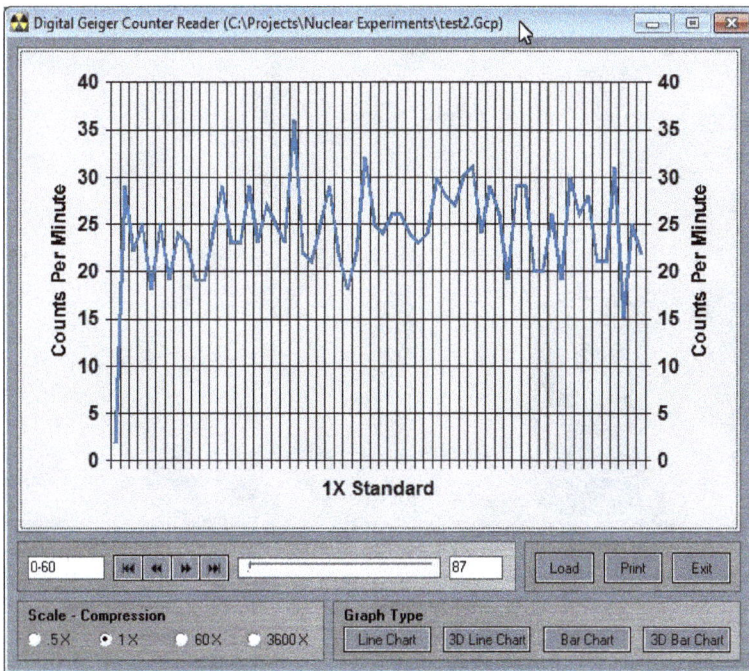

Figure 8.3

Next, we take the wand and place it close to the KCl salt. You can lay the wand flat in a tray filled with salt, as shown below.

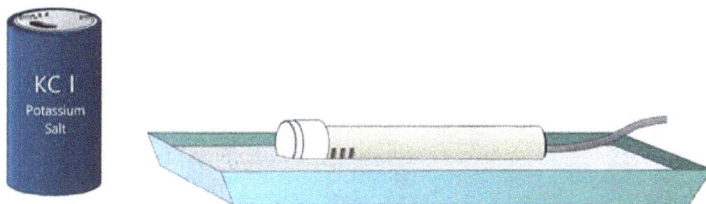

Figure 8.4

Alternatively, you can also place the wand in a beaker filled with KCl salt, as shown below.

Figure 8.5

The radiation readings have more than doubled in comparison to the background radiation readings.

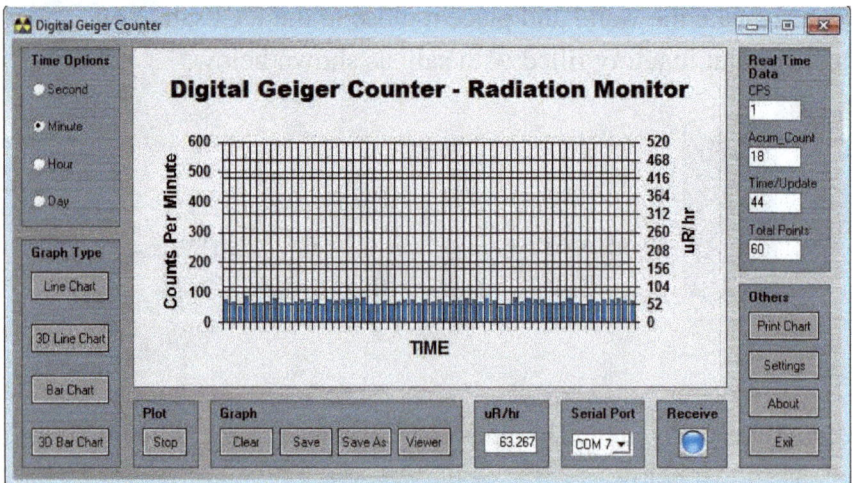

Figure 8.6

Again, we can save the data and open it up in the viewer program.

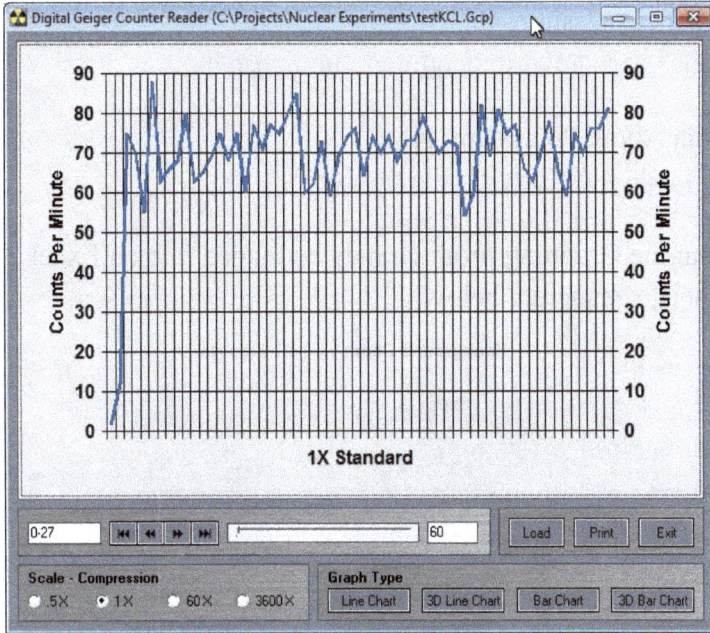

Figure 8.7

Potassium-40 (K-40) is a naturally occurring radioactive isotope. Natural potassium contains 0.0118% of the K-40 isotope. The K40 isotope has a half-life of 1.28 billion years. Output radiation consisting of 89.3% beta (mean energy of 560 keV and beta maximum energy is 1.31 MeV) and 10.7% gamma with 1.461 MeV.

This information may be used for further experiments.

1. Confirm that the excessive radiation detected from the KCl does not consist of alpha radiation. A simple paper shield should suffice.
2. Confirm that approximately 89% of the excessive radiation from the KCl consists of beta radiation.

More About Potassium-40:
https://pubs.usgs.gov/gip/geotime/radiometric.html

Wikipedia About Potassium-40:
https://en.wikipedia.org/wiki/Potassium-40

Data from the Graphing program may be loaded into an Excel spreadsheet, see picture below.

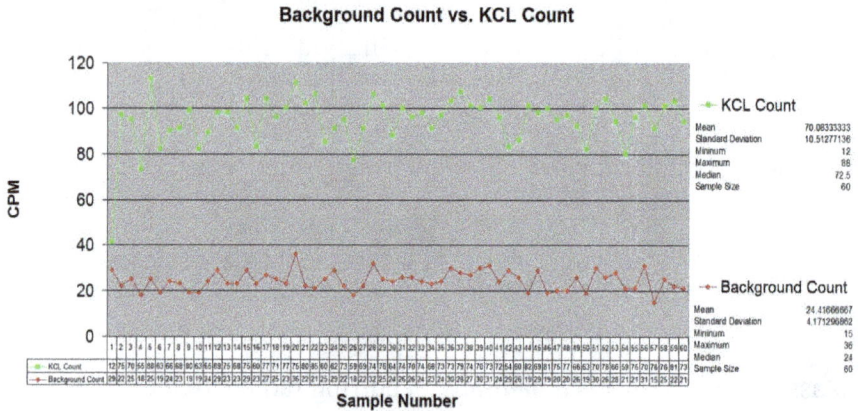

Background Count vs. KCL Count

Figure 8.8

The red line shows the background radiation count and the green line details the increased radiation count from the KCl salt. Excel allows one to easily determine the Mean, Median, Min, and Max CPM for the samples as shown at the side of the table.

Chapter 9

9 - Capturing and Detecting Radon in the Environment

Radon nuclides captured in ambient air by electrostatically charged balloon

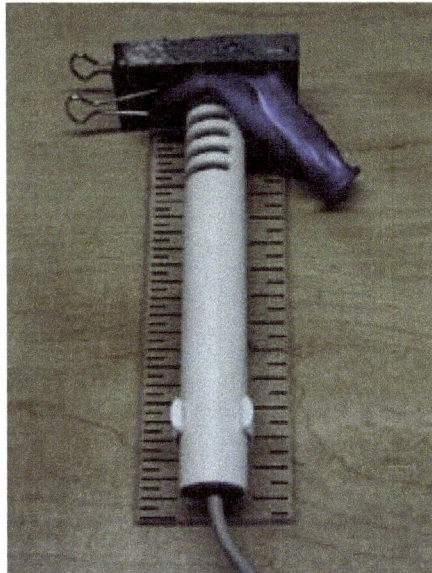

Figure 9.1

A child's balloon can be used to make an impressive demonstration. When the balloon is inflated, an electrostatic charge is placed on the balloon by rubbing the balloon against one's hair, sweater, or a microfiber cloth.

Figure 9.2

The electrostatically charged balloon is suspended in an indoor environment for 60 minutes.

Figure 9.3

The charged balloon will accumulate radioactive atoms from the decay of radon in the ambient air making the balloon radioactive. The balloon's radioactivity will increase 10X (or greater) the normal background radiation as detected by a Geiger counter. The Geiger counter must be sensitive to alpha particles to measure the full increase in radioactivity. The origin of this experiment is based on a 1995 paper by Walkiewicz titled "The Hot Balloon (Not Air)."

Radon is a noble gas. The half-life of radon is 3.82 days. As radon decays, the daughter products of radon become attached to positive charged aerosol particles. These positive charged particles are attracted to and captured by the negative charged balloon. The decay of Radon gas follows. Look at this information below, we can see the majority of radioactivity is confined from the decay in the first four steps. The decay of ^{210}Pb takes 22.3 years and can be ignored for our purposes here.

^{222}Rn, 3.8 days, alpha decaying to...
^{218}Po, 3.10 minutes, alpha decaying to...
^{214}Pb, 26.8 minutes, beta decaying to...
^{214}Bi, 19.9 minutes, beta decaying to...
^{214}Po, 0.1643 microseconds, alpha decaying to...
^{210}Pb, which has a much longer half-life of 22.3 years, beta decaying to...
^{210}Bi, 5.013 days, beta decaying to...
^{210}Po, 138.376 days, alpha decaying to...
^{206}Pb, stable.

We do not know in what proportion the balloon has accumulated the various daughter nuclides. Therefore, the half-life of the radioactivity captured on the balloon will vary. It is safe to estimate the half-life of the balloon's radioactivity will be between 30 and 40 minutes. If we look at the decay of ^{218}Po, its half-life is only 3.1 minutes. In 21 minutes, its contribution to the radioactivity will be 1/128 of its original amount. The Geiger counter's sensitivity to the various energies of the decay radiation also plays a part. From the Radon decay sequence shown above, the majority of the radiation is beta and alpha radiation.

Doing the Experiment

Inflate the balloon. I tied the end of the balloon with a wire tie to seal it. I used a wire tie because it allows me to remove the tie and gently deflate the balloon. One doesn't want to knock off all the radioactive nuclei on the balloon's surface by popping the balloon.

Figure 9.4

The balloon remains suspended for about 60 minutes. After the 60 minutes has elapsed, the balloon is retrieved and gently deflated. The limp balloon is placed next to the alpha window of the GM tube. I used the GCA-07W digital Geiger counter to measure the radioactivity.

Figure 9.5

The following picture shows the graphing software response from background radiation to when the balloon is placed in front of the GM tube. The software graph shows about one hour of data.

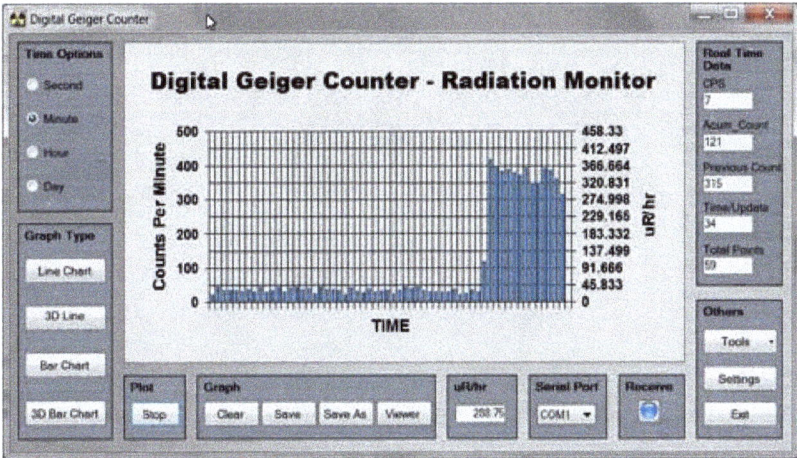

Figure 9.6

The data from 6 hours of the software readings were saved and then exported into a CVS file that was loaded into an Excel spreadsheet and graphed, see below.

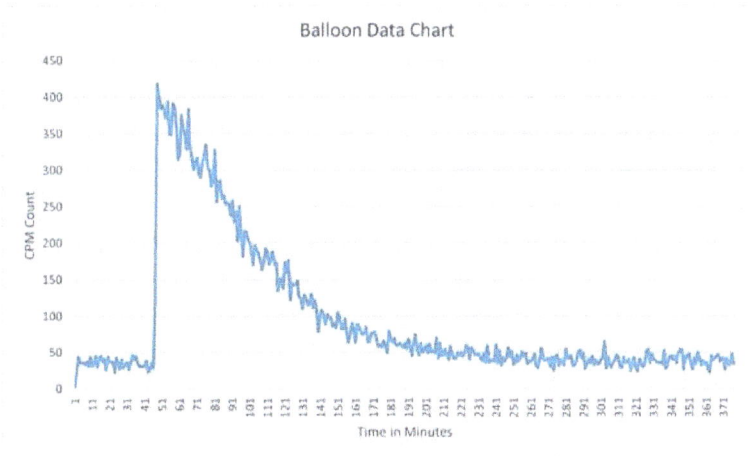

Figure 9.7

Chapter 10

10 - Calculating Dead Time of GM Tube

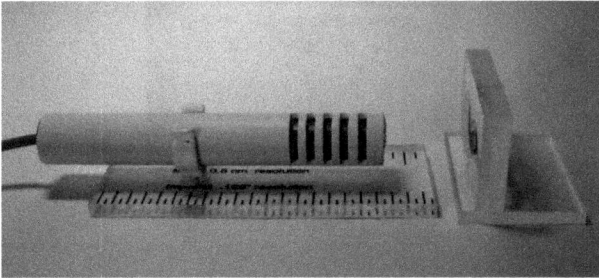

Figure 10.1

A GM tube detects radiation when the radiation (particle or ray) ionizes the gas inside the GM tube. This ionization initiates an avalanche of electrons traveling to the anode and positive ions toward the cathode. Upon reaching the anode, the avalanche creates a voltage pulse. That pulse is detected by the Geiger counter electronics, and we typically hear this as an audible click from the Geiger counter.

During the detection process described above, if another particle or ray ionizes more gas inside the GM tube, would the GM tube see it as a multiple pulse? The answer is no. During the time of detection, the GM tube cannot detect any additional radiation until the GM tube has finished detection and has a chance to reset.

The reset time for the GM tube is also called dead time, because during this detection-reset time, the GM tube is "dead" and can't detect radiation. The dead time for GM tubes varies from tube to tube.

The loss of counts due to dead time can be important for some measurements that require a high degree of accuracy. In this experiment, we will calculate the dead time of the GM tube using a two-source method.

There are specialty sources for this procedure that look like a standard source that has been cut in half. We are not using specialty sources, instead, we are using two 10 uCi CS-137 sources. Basically, we have a holder that can hold two CS-137 1" diameter x 1/8" thick discs next to one another.

Figure 10.2

We take three readings.

The first reading is from source one s_1. The second reading is both sources s_1 and s_2. The third reading is from s_2.

Figure 10.3

The source s_1 is placed at a distance from the wand to obtain an approximate CPM reading of 15,000 to 20,000 Counts Per Minute (CPM).

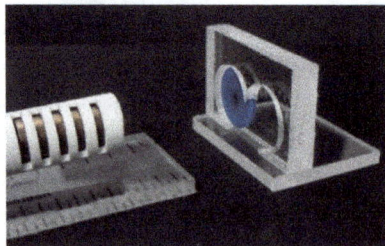

Figure 10.4

Once this distance has been established, the position held is the same for all three readings.

Run # 1			
	CPM Count S_1	CPM Count S_{12}	CPM Count S_2
Reading	15628 CPM	37372 CPM	23818 CPM
Divide by 60	/60	/60	/60
Reading in CPS	260 CPS	622 CPS	396 CPS

Deadtime Equation

$$T = \frac{S_1 + S_2 - S_{12}}{2 s_1 s_2}$$

Putting the numbers in

$$T = \frac{260 + 396 - 622}{2(260)(396)}$$

Results

$$T = \frac{34}{205920}$$

Deadtime

$$T = 0.000165$$

Deadtime in usec

$$T = 165 \ usec$$

Performing the experiment

When running the experiment, we recommend you take five reading for s_1, s_2 and s_{12}, and take the average of the readings.

Run #			
	CPM Count S_1	CPM Count S_{12}	CPM Count S_2
Reading			
Divide by 60	/60	/60	/60
Reading in CPS			

BACKGROUND _____ CPM

Chapter 11

Chapter II

11 - Backscattering Experiment

This experiment investigates the deflection (scattering) of radiation by matter. Radiation given off by a source is emitted in all directions surrounding the source. The portion of that radiation that passes through a GM tube has a possibility of being detected. Depending upon the efficiency of the GM tube, on average the GM tube will detect about 10% of the radiation entering the GM tube.

In this experiment, we will use beta particles from a SR-90 source. Strontium 90 (^{90}Sr) has a half-life of 28.79 years. It undergoes a beta decay to yttrium (^{90}Y) with the release of a beta particle of 0.546 MeV. Yttrium has a half-life of about 64 hours with a beta decay to ^{90}Zr (zirconium), with the release of a beta particle of 2.28 MeV. The zirconium is stable.

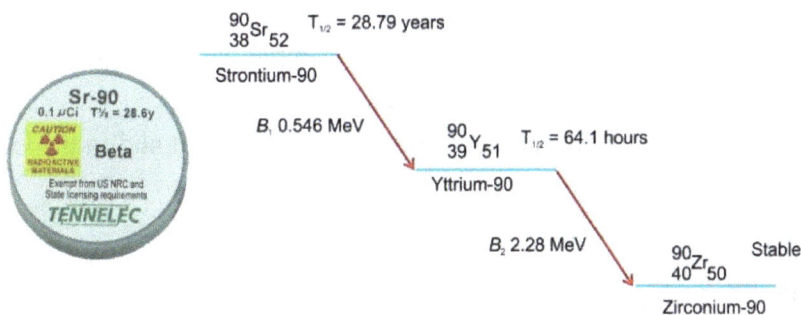

Figure 11.1

When a beta particle enters matter, its path can be deflected. If a beta particle is deflected a full 180 degrees from its initial direction upon entering the material, it is called backscattering.

The geometry of the experiment is arranged so that beta radiation deflected by a sample block is counted by the GM tube. The shield in-

107

between the source and GM tube blocks the radiation directly from the source to the GM tube.

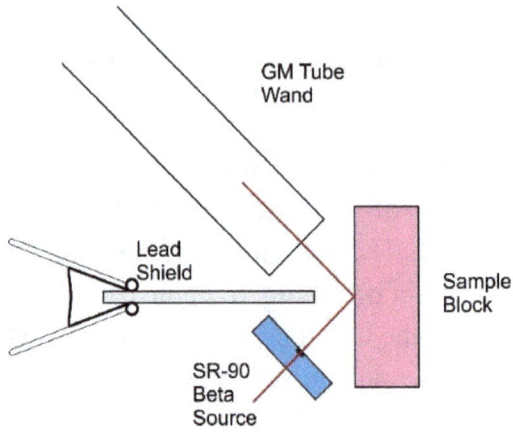

GM Tube
Wand

Lead
Shield

Sample
Block

SR-90
Beta
Source

Figure 11.2

In a perfect experiment, the initial reading would take place in a vacuum. That is impractical for our experiment, so we start the experiment in free air. There are a few factors that influence backscattering. Among them are the following:

- Energy of the beta particle
- Atomic Number (Density) of the sample block material
- Thickness of the blocking material
- Distance of the source to the blocking material and to the GM tube

As the pictures show below, the geometry is pretty tight; however, using this tight geometry produced good results.

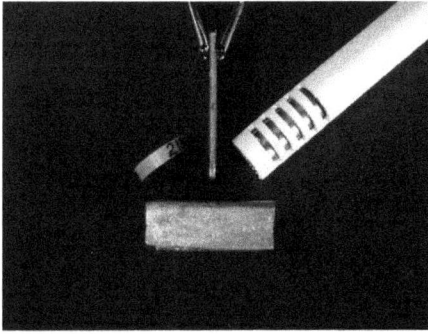
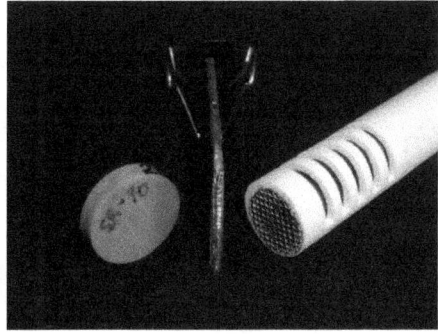

Figure 11.3   *Figure 11.4*

In this experiment, we used three different sample block materials: aluminium, iron, and lead.

Blocking Material	Atomic Number	Count CPM
None (air)		47
Aluminum	13	102
Iron	26	125
Lead	82	207

NOTES:

When performing this experiment, you should take five CPM readings for each sample and average your results. In addition, take a background reading before beginning the experiment and subtract the background readings to improve the accuracy of your results.

Background _____ CPM

Backing Material	Atomic Number	Count CPM	CPM - Background Corrected Count

Alternative Experiment

This alternative experiment is easier to perform. A small piece of 1/16" thick lead approximated 2" x 3.25" is wrapped around a 1" diameter cylinder to fashion it into a tube.

A wand holder suspends the wand approximately 2-1/4" above the SR-90 beta source. CPM readings are taken from the beta source. Next, the lead tube is placed on top of the SR-90 beta source and the measurements are repeated. Count without lead tube was 273 CPM. Count with lead tube was 592 CPM.

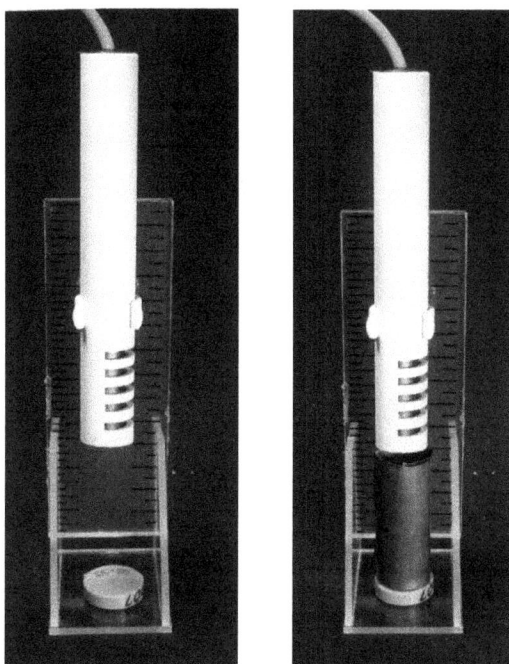

Figure 11.5

Chapter 12

12 - Magnetic Deflection of Beta Particles Experiment

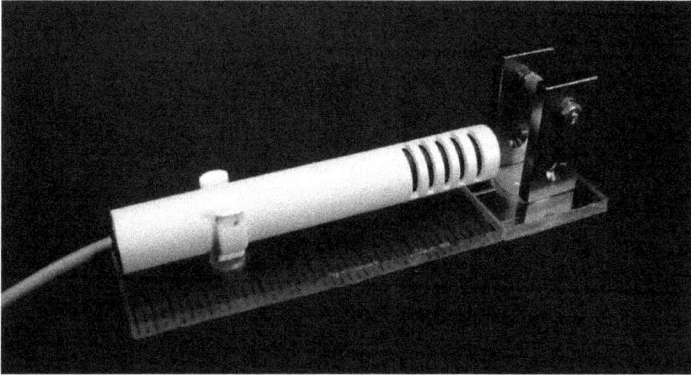

Figure 12.1

Beta particles and electrons are the same particle. The reason we classify them differently is that the beta particle originates from the nucleus of the atom being generated by radioactive decay. So, we could say that a beta particle is a freshly minted electron. If we place a strong magnetic field across the path of beta particles being emitted from a radioactive source, we can make the path of the beta particles bend. The energy of the beta particles and the strength of the magnetic field will determine the degree of deflection.

The following graph provides an illustration of the deflection of charged particles in a magnetic field. Notice the gamma ray photon is undeflected by a magnetic field.

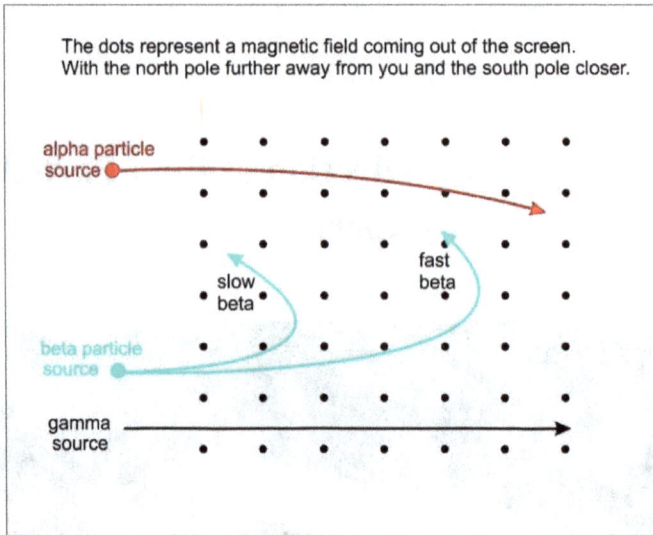

The dots represent a magnetic field coming out of the screen.
With the north pole further away from you and the south pole closer.

alpha particle source

slow beta

fast beta

beta particle source

gamma source

Figure 12.2

The objective of this experiment is to deflect beta particles using a magnetic field and to detect the deflection using a Geiger counter.

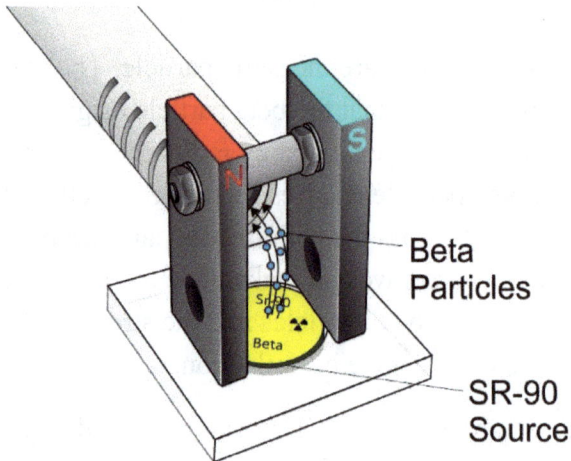

Beta Particles

SR-90 Source

Figure 12.3

Fabricating the Magnet:

I do not have a horseshoe magnet strong enough for this experiment. So, I fabricated an equivalent horseshoe magnet out of two very strong neodymium magnets and a little hardware, see exploded view below.

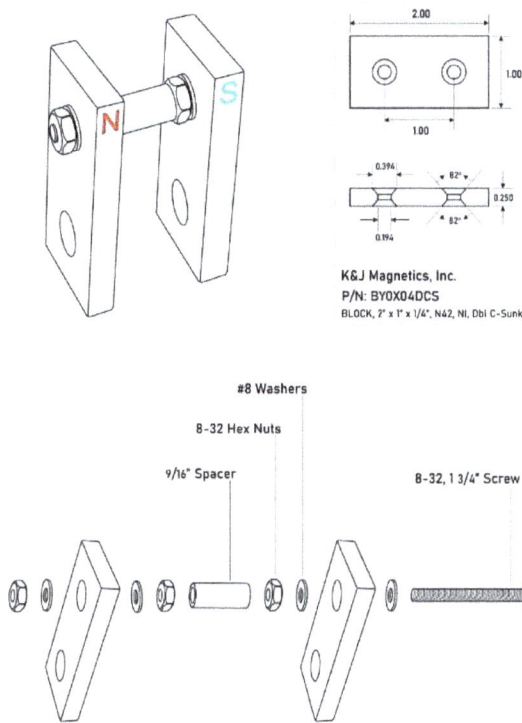

K&J Magnetics, Inc.
P/N: BY0X04DCS
BLOCK, 2" x 1" x 1/4", N42, NI, Dbl C-Sunk

#8 Washers

8-32 Hex Nuts

9/16" Spacer

8-32, 1 3/4" Screw

Figure 12.4

The magnets are from K&J Magnetics. The base is made from a 2" square by ¼" thick piece of acrylic with a one-inch hole drilled through it. The 1" hole is for fitting the SR-90 source. That base makes it easier to rest the magnet.

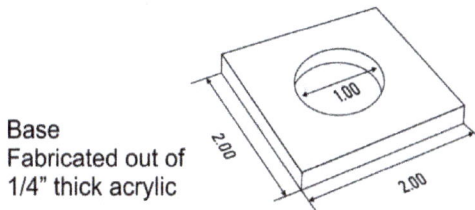

Base
Fabricated out of
1/4" thick acrylic

Figure 12.5

117

Performing the Experiment:

Step 1) Begin by using the Geiger counter to take a set of background radiation readings. Set GCA-07W to CPM mode using the front panel switch. Log a minimum of three one-minute readings. If you are using the radiation graphing program included with the CGA-07(W), start the program. Set the time increment to one minute.

Step 2) Next, place the SR-90 source in the base by the detector, with no magnetic field. Take a minimum of three one-minute recordings. The position of the SR-90 source as it relates to the GM tube detector is critical. It should be at the same distance when placing the magnet on top of the SR-90 source and base.

Failure to maintain this exact distance will render the background numbers inaccurate for the experiment.

Step 3) Place the magnet on top of the SR-90 source in position by the detector, with N-S magnetic field. Take a minimum of three one-minute recordings.

Step 4) Lift the magnet and rotate it 180 degrees and place it back on the SR-90 source in position by the detector, with S-N magnetic field. Take a minimum of three one-minute recordings.

In steps 2, 3, and 4, the distance between the SR-90 source and detector should remain the same, or as close as possible to record proper data.

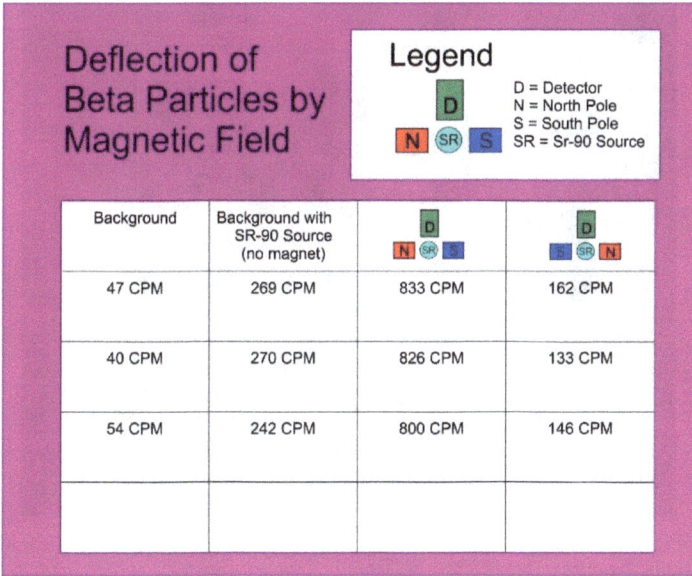

Deflection of Beta Particles by Magnetic Field

Legend

D = Detector
N = North Pole
S = South Pole
SR = Sr-90 Source

Background	Background with SR-90 Source (no magnet)	D / N SR S	D / S SR N
47 CPM	269 CPM	833 CPM	162 CPM
40 CPM	270 CPM	826 CPM	133 CPM
54 CPM	242 CPM	800 CPM	146 CPM

Images Radiation Graphing program allows you to save data and reload it into its own data viewer program.

The wand holder shown in the first picture in this chapter is the Images SI sled holder for the GCA-07W wand.

A close-up of the experiment is shown in the following picture.

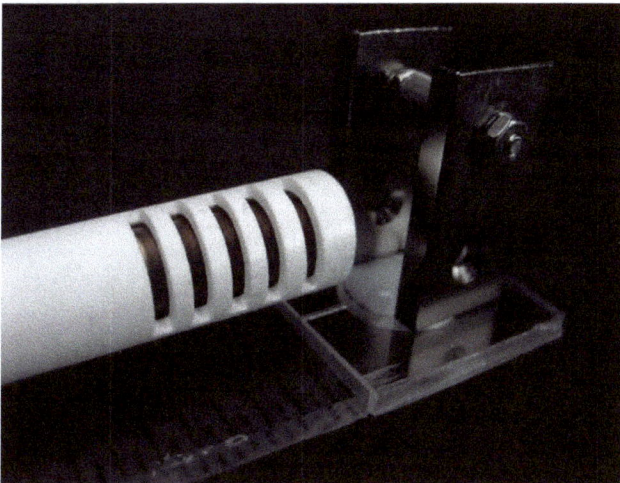

Figure 12.6

119

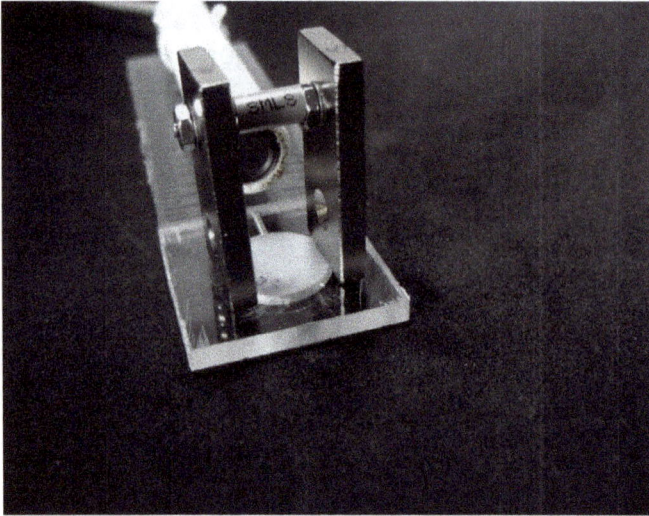

Figure 12.7

The results of the experiment show that beta particles are deflected by the magnetic field.

If we outline the magnetic field as it is applied in the experiment, when it is directing beta particles toward the detector, it would look like the following illustration.

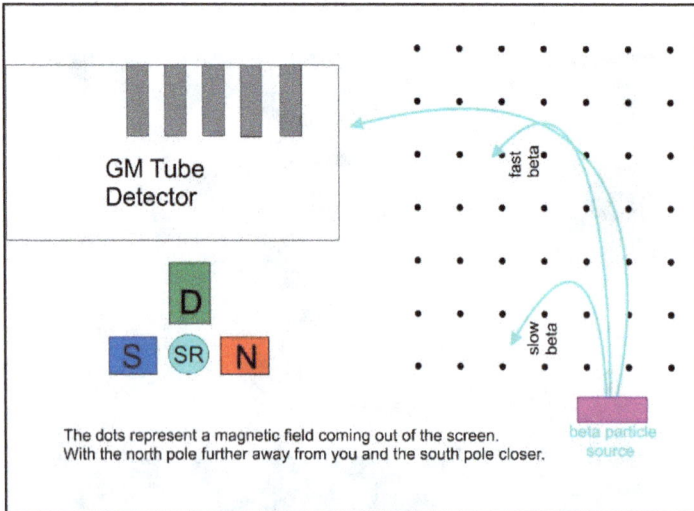

Figure 12.8

Going Further

If we changed our Sr-90 source to a Na-22 source that emits positive beta particles (Positrons), what would you expect the deflection of the positrons to be, compared to the normal beta particles?

Chapter 13

13 - Half-Life Experiment

This experiment examines radioactive decay. The radioactivity of a material is in proportion to the number of atoms that decay and emit radioactive particles in an interval of time.

This is the formula for the radioactivity of a material with respect to time.

$$A_t = A_0 e^{-\lambda t}$$

The radioactivity of the material A_t at any time is equal to A_0 (initial activity) multiplied by e raised to the power of negative λ lambda times t. Lambda λ is the decay constant of the material.

The Half-life of a radioactive material (symbol t ½) is the time it takes for its radioactivity to reduce to half of its initial value. We will calculate the half-life of a radioactive isotope.

To perform this experiment, you will need a Cs-137/Ba-137m - Isotope Generator Kit from Spectrum Techniques company.

http://www.spectrumtechniques.com/products/sources/isotope-generator-kit/

How Isotope Generator Works

Cesium-137 is a radioactive material with a half-life of 30 years. When Cs-137 decays, with the emission of a beta particle, it becomes either meta-stable Ba-137m (94.6%; beta energy 0.514 Mev) or stable Ba-137 (5.4% beta energy 1.174 MeV). The meta-stable Ba-137m has a half-life of 2.55 minutes; it decays to stable Ba-137 with the emission of a gamma with the energy of 0.66 MeV.

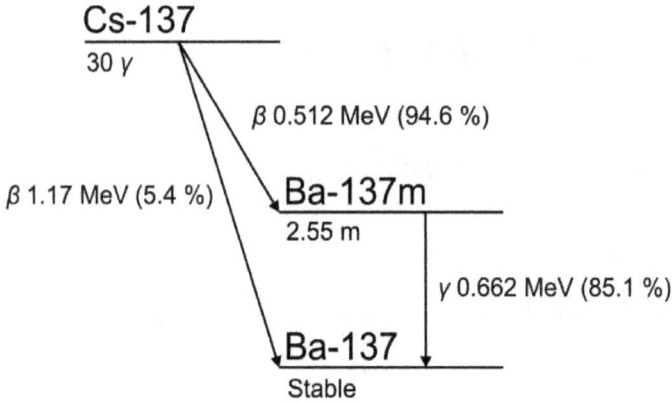

Cs-137
30 y

β 0.512 MeV (94.6 %)

β 1.17 MeV (5.4 %)

Ba-137m
2.55 m

γ 0.662 MeV (85.1 %)

Ba-137
Stable

The isotope generator contains an exempt quantity of Cs-137, in the form of a cesium salt, CsCl. As the cesium decays to meta-stable Ba-137m, the Ba-137 has different chemical properties than the cesium salt. These chemical properties allow the Ba-137m to be extracted from the Cs-137 using a weak acidic salt solution, called the eluting solution.

By passing the eluting solution through the generator, we obtain a solution containing the meta-stable Ba-137m that we can use for our experiment. Passing the eluting solution through the generator is sometimes referred to as "milking" the generator. Once the Ba-137m has been washed out of the generator, the Cs-137 is left behind to regenerate the Ba-137m. The regeneration of Ba-137 will occur in less than one hour.

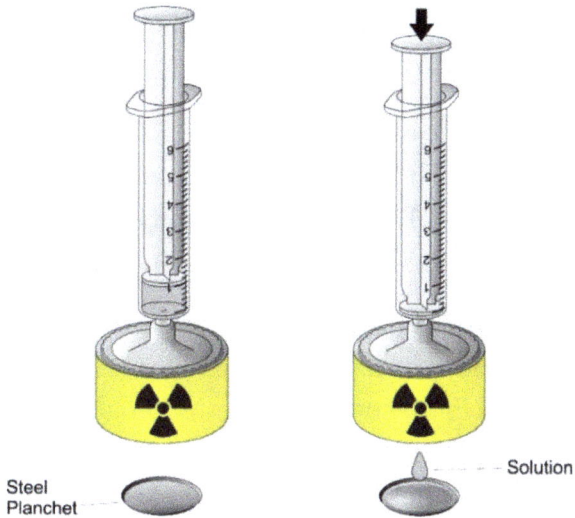

Figure 13.1

All the materials necessary to obtain a meta-stable Ba-137 solution is included in the kit. Follow the directions included with the kit.

http://www.spectrumtechniques.com/wp-content/uploads/2016/12/Isotope-Generator-Brochure.pdf

After you have filled the steel planchet with the meta-stable Ba-137m solution, it is placed under the wand for taking measurements.

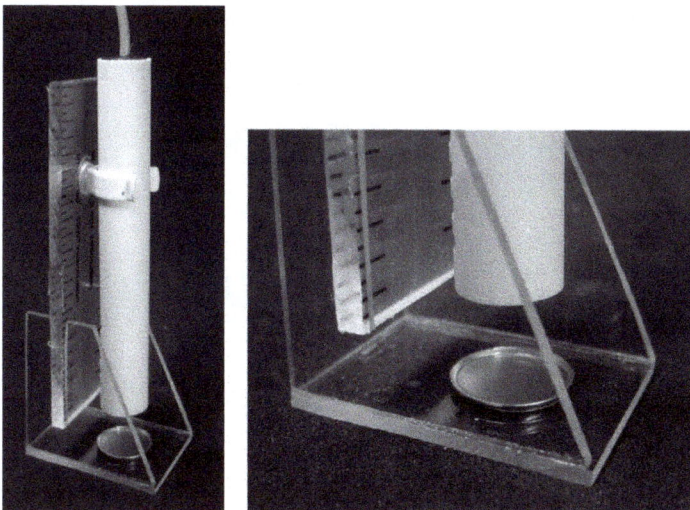

Figure 13.2

For this experiment, I used the Geiger graphing software that is included with the GCA-07W.

To calculate the half-life of meta-stable Ba-137, we must first find its decay rate λ lambda. To find lambda, we need to know how the radioactivity of our Ba-137m changed over time. I set the time base to one second and recorded 330 one-second samples (5-1/2minutes).

Once you have recorded this time interval, save the recording in a file. Once this file is saved, you can perform a number of operations. You can view the sample in the software's viewer program. Convert the file to a .csv file to be used in a spreadsheet. First convert the file to a csv file. Next, view the csv file with Notepad to see all the record one-second samples in the file. To see how the activity changed over time, we look at two numbers from the 330-second sample interval, the first sample, and the last sample. Record these two numbers. For my sample, the first number was 134 and the last sample recorded was 32.

We use the following equation to calculate lambda.

$$\lambda = \frac{-\ln(\frac{Ao}{At})}{t}$$

Now we just need to plug in our numbers.

$$\lambda = \frac{-\ln(\frac{32}{134})}{330}$$

Do the division

$$\lambda = \frac{-\ln(.239)}{330}$$

If you do not have a calculator that can calculate the natural log of 0.239, go to the following online calculator:

https://www.rapidtables.com/calc/math/Ln_Calc.html

Calculate the natural logarithm of 0.239.

$$\lambda = \frac{-1.43}{330}$$

Finish up the math.

$$\lambda = 0.00433s^{-1}$$

Now we have the decay constant. We can use this to calculate the half-life.

$$T_{\frac{1}{2}} = \frac{\ln(2)}{\lambda}$$

Go to the natural log calculator and find the natural log of 2, which is 0.693. Populate the equation with numbers.

$$T_{\frac{1}{2}} = \frac{.693}{.00433}$$

Perform the math

$$T_{\frac{1}{2}} = 160s$$

This answer is in seconds. To convert to minutes, divide the answer by 60.

$$T_{\frac{1}{2}} = 2.66m$$

The textbook half-life of Ba-137 is quoted at 2.55 minutes. Our calculation of 2.66 comes pretty close.

The exported data from the Geiger Graphing Program (the csv file we created) may be imported into Excel. Once in Excel, we can graph the data.

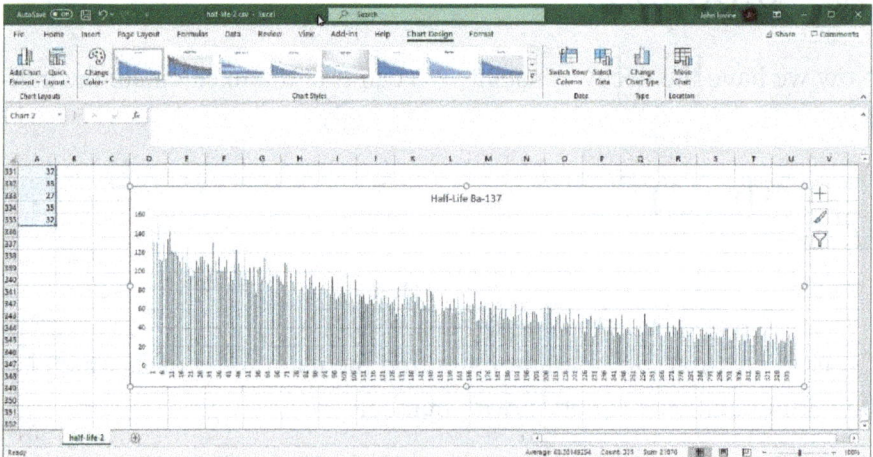

Figure 13.3

The graphic plot, while useful, is a little dirty with the raw data points. So, I did another run in CPM mode for 15 minutes. The graph below is the raw data on the Geiger Graphing program.

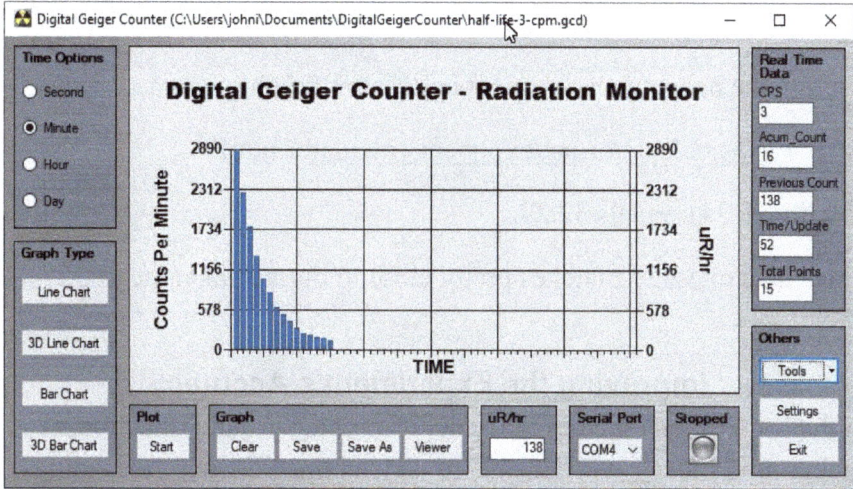

Figure 13.4

The first count is 2890 CPM and the last count is 138 CPM. Try running the equation with these numbers. Remember the original numbers were CPS, these numbers are CPM.

Last Check

We have our half-life answer, why don't we plug our numbers back into our original equation and see how close we come to our last reading at 330 seconds.

$$A_t = A_0 e^{-\lambda t}$$

Here we want to calculate the value of A_t at 330 seconds. We know our initial value A_0 is 134, e is Euler's number (approximately 2.718) raised to the decay constant lambda λ times t at 330 seconds. The decay constant is 0.0043 times 330 which equals -1.42. So, we raise e (2.718) to an exponent value of -1.42. This calculates to a value of 0.24171.

For anyone needing an exponent calculator, go online to:

https://www.calculator.net/exponent-calculator.html?

134 time 0.241 equals 32.29.

This value of 32.29 matches pretty close to our actual value of 32.

Improving the Experiment's Accuracy

In this experiment, we did not account for background radiation. If background radiation is considered and subtracted from the readings, we will improve the accuracy of our results.

Chapter 14

14 - Urban Prospecting

You live far from the uranium mining in New Mexico, Utah, and Colorado, so where can you go prospecting for radioactive materials? You may be surprised to learn radioactive materials are readily available, even in urban areas.

If you go to a well-stocked hardware store, you could find Coleman Lamp Mantles. Coleman lamp mantles are manufactured using Thorium Dioxide. If you check the mantle with a Geiger counter that can detect alpha particles, you will find the mantle is radioactive due to the radioactivity of the Thorium which releases a radioactive gas radon-220.

The amount of radioactivity given off by the mantle was studied in the early 1980s and has been considered safe since 1981.

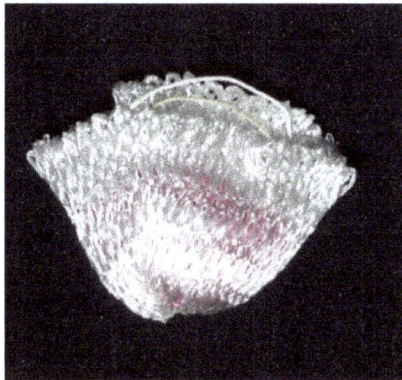

Figure 14.1

However, the newer mantles made by Coleman are no longer radioactive. Sometimes, mantles manufactured overseas for the Coleman Lanterns are still radioactive.

An alternative is to look for vintage Coleman mantles on Ebay.com

Flea markets, Swap Meets, and Ebay.com

What you can find in flea markets, swap meets, and searching ebay.com are the following:

- Fiesta Ware
- Radium Clocks and dials
- Uranium Glass
- Uranium ore specimens
- Ion Chambers
- Quack Radioactive Health Products from the Early 1900's

Fiesta Ware

Figure 14.2

Fiesta dinnerware was introduced in 1936. It was manufactured in five colors. The red glaze color was the most radioactive because of the quantity of uranium oxide added to the glaze to achieve the red/orange color. Interestingly, during World War II, the U.S. government confiscated the company's uranium materials. So red Fiesta ware disappeared until 1959. The Fiesta ware produced in these later years used depleted uranium (DU).

The Fiesta ware products were discontinued in 1969. However, one should not discount other red ceramic glaze products from the same era. Other manufacturers also used uranium in their glaze compounds.

Radium Clocks, Dials, and Pocket Watches

There are old-style 'glow in the dark' clocks and watches that use radium paint on their numbers and dials. To detect the radiation from the radium paint, the glass front must be removed, and the alpha sensitive GM tube or wand held close to the paint.

Figure 14.3

Uranium Glass

Uranium Glass is also called "Depression Glass" and "Vaseline" glass. Depression glass was made during the 1920s and 1930s. As with Fiesta ware, this glass had a uranium oxide added for coloration. The color of the glass ranges from a translucent pale yellow (Vaseline color) to transparent green. The glass fluoresces bright green under ultraviolet light. Radiation is minimal, and registers slightly above background radiation. The pictures below are of uranium glass marbles and a uranium glass flower frog.

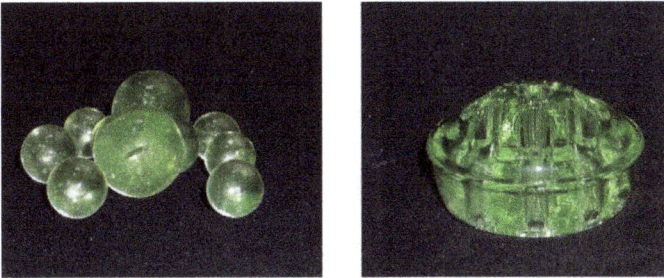

Figure 14.4

Uranium Ore and Radioactive Minerals

You can purchase uranium ore on Amazon.com or eBay.

Figure 14.5

https://www.amazon.com/Images-SI-Uranium-Ore/dp/B000796XXM

Ion Chambers

Some smoke detectors use a small amount of radioactive material in their ion chamber.

Figure 14.6

Quack Radioactive Health Products

I have collected two such items. The first are radium pills, see pictures below.

Figure 14.7

The next item is the Revigator. This is a ceramic water containment vessel lined with radioactive carnotite material. The intention was to put clean water into the Revigator and let the water sit overnight. This allowed the water irradiated by the uranium and infused with radon gas from the lining material, making the water radioactive. At which point the water is consumed with the belief that it will improve health and prevent things like arthritis and senility.

Figure 14.8

Good hunting.

Chapter 15

15 - DIY Geiger Counter

DIY Build Your Own Geiger Counter - Radiation Detector

Figure 15.1

This chapter shows you how to build a fully functional Geiger counter capable of measuring the three primary forms of radiation: alpha, beta, and gamma radiation. The Geiger counter is sensitive enough to detect background radiation. It's expandable. You can enhance the basic Geiger counter by adding a Digital Meter Adapter (DMAD) that adds a digital output for the Counts Per Second (CPS). The DMAD can be connected to a Windows PC that is running Images SI Inc's free Windows Radiation graphing software. Connecting the DMAD to the PC requires a USB TTL adapter. The DMAD also has a true Random Number Generator function. We will talk more about the DMAD in the next chapter.

The Windows Radiation Graphing Program is free and available for downloading at https://www.imagesco.com/geiger/geiger-graph.html

145

The Geiger counter produces an audible click and blinks a LED each time a radioactive particle is detected. It has a Data output jack, that outputs a +5V pulse every time a radioactive particle is detected. It also has a headphone jack for private listening. Typically, the Geiger counter will click randomly about 20 times a minute due to normal background radiation.

Geiger Counter Schematic

You can purchase the components and build this circuit on a prototyping breadboard, or you can purchase a kit. This chapter is on building the kit.

The circuit is shown in Figure 15.2. The 4049 Hex Inverting Buffer is set up as a square wave generator. The power MOSFET IRF830 switches the current on and off to the primary windings of the mini step-up transformer. The output of the mini step-up transformer is fed to a voltage doubler, consisting of two high voltage diodes D4 and D5, and two high voltage capacitors C3 and C4.

The high voltage output from this stage is regulated to 500 volts needed for our GM tube, *GMT-01 (LND 712), by three Zener diodes stacked one on top of the other (D6, D7 and D8). Diodes D7 and D8 are 200V Zener diodes and diode D6 is a 150-Volt Zener. Together (200 + 200 + 150 = 550), they equal 550 volts. Five hundred fifty volts is the optimum operating voltage for our GM Tube.

The 500-volt regulated output is fed to the anode of the LND 712 GM tube through a current limiting 10 mega-ohm resistor R16. The 10 mega-ohm resistor limits the current through the GM tube and helps quench the avalanched ionization when a radioactive particle is detected.

146

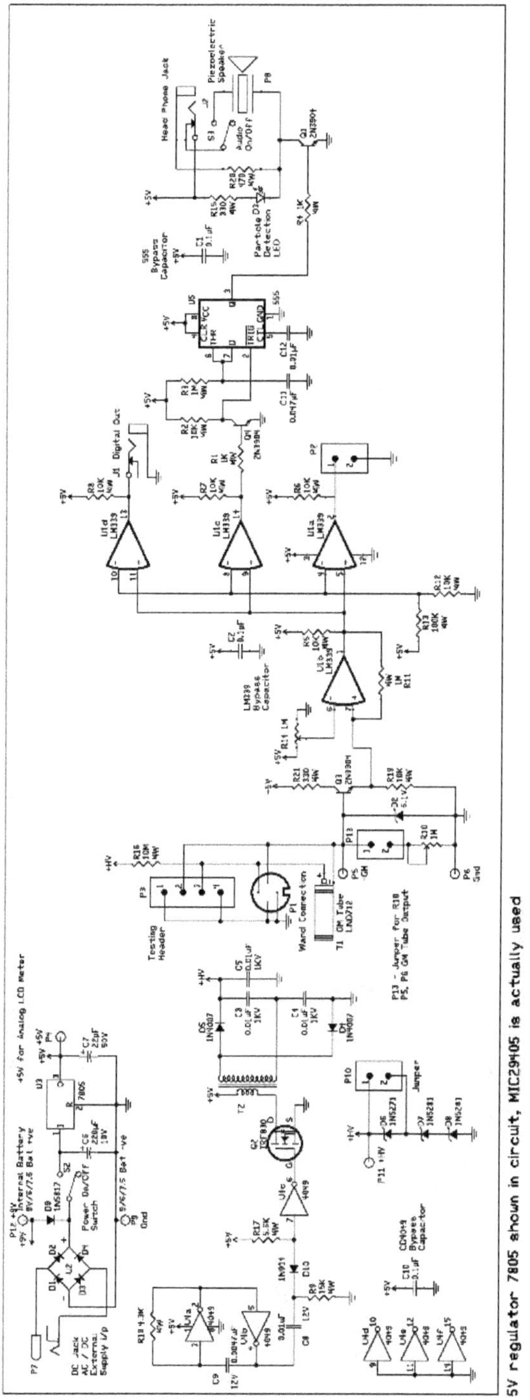

Figure 15.2

147

5V regulator 7805 shown in circuit, MIC29405 is actually used

*To regulate to the 400 volts needed for the GMT-02 Tube, a jumper is placed on P10. This jumps the 100-volt Zener diode at D6. R16 becomes a 2.2 mega-ohm resistor when using the GMT-02.

The cathode of the tube is connected to a 5.1V (D2) Zener diode. The voltage pulse across D2 generated by the detection of radiation, feeds to the base of a 2N3904 NPN transistor.

The NPN transistor clamps the output pulse from the GM tube to Vcc and feeds it to a comparator gate on the LM339. The pulse signal from the gate, pin 14 of the LM339, is a trigger to the 555 Timer through Q4. The timer is set up in monostable mode that stretches out the pulse received on its trigger. The output pulse from the timer flashes the LED and outputs an audible click to the speaker via pin 3.

You may hardwire this circuit to a breadboard or use the available PCB board. Although you do not need the PCB, the PCB will make construction easier.

PC Board Construction

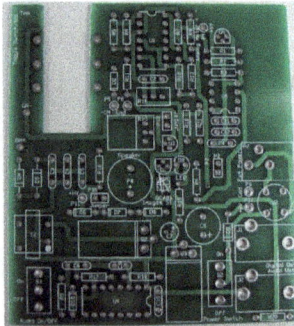

Figure 15.3A. Detail of top silkscreen on PCB (for GMT-01 or internal tube)

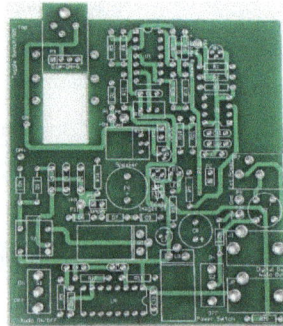

Figure 15.3B. Detail of top silkscreen on PCB (for GMT-02 or external wand)

The top silkscreen of the PCB is shown in Figure 15.3. Begin construction by soldering resistors R17 5.6K (color bands green, blue, red), R18 4.3K (color bands yellow, orange, red) and R9 15K (color bands brown, green, orange). Next, we will wire the square wave

148

generator and pulse shaping circuit using the ICS-16 socket for the 4049, marked U4 on PCB. Insert the ICS-16, making sure to orient the notch on socket to the drawing on the PCB and solder to the board.

Place and solder components C8 (.01uf), C9 (.0047uf), C10, (.1uf) and D10 (1N914). Now construct the high voltage section consisting of the step-up transformer T2, diodes D4 & D5 (1N4007), and capacitors C3, C4 and C5 (.01uf 1KV). Mount IRF830 transistor Q2 to the PCB, bend the transistor outward so it lays flat on the PCB, see Figure 6, and solder.

To this, add the 5-volt 7805 regulator (U3), bending it outward so it lays flat like the transistor and solder into position. Next, mount and solder capacitors C6 (220uf-330uf), C7 (22uf), and diode D9 (1N5817). Place and solder the 9-volt battery cap on the PC board. The red lead connects to the positive terminal P12. The black lead connects to GND, marked P9. Solder the power switch to the PCB at S2. Insert 4049 into the socket, making sure to orient the notch on the chip to the notch on the socket.

Testing HV Section

CAUTION: Circuit generates high voltage power that can provide an electrical shock. Exercise caution when working around the high voltage section of the circuit. The capacitors C4 and C5 can hold a HV charge after the circuit has been shut off.

To check the HV power supply; turn the power switch off. Insert the 9-volt battery onto the battery cap. Set up a VOM to read 500 to 1000 volts. Place the positive lead of the VOM at P11. The negative lead of the VOM is connected to the—(negative) terminal of the 9-Volt battery.

Apply power to the circuit using the power switch. The circuit should generate anywhere between 550 and 800 volts (depending upon component tolerances). If you are reading between 550 and 800 volts, fine, turn off the power. Add the three Zener diodes; D6 (150V), D7 & D8 (200V). Attach a 2-pin header and jumper at P10. Apply power again, with the positive lead of the VOM still attached to P11; you should read a voltage of 380-400 volts. If you're not getting a proper voltage reading, check the Zener diodes to make sure you have them orientated in the right direction.

Figure 15.4. Completed HV section of the circuit

Installing GM tube - Continuing Construction

Finish the construction of the circuit by adding the ICS-8 for the 555 timer and the ICS-14 for the LM339. Mount and solder all remaining resistors; R1 is a 1K resistor (color bands brown, black, red). R2, R4-R8, R12 & R19 are 10K resistors (color bands brown, black, orange). R3 & R11 are 1 Meg resistors (color bands brown, black, green). R13 is a 100K resistor (color bands brown, black, yellow). R15 & R21 are 330-ohm resistors (color bands orange, orange, brown). R16* is a 2.2 Meg resistor (color bands red, red, blue), and R20 is a 470-ohm resistor (color bands yellow, purple, brown).

150

Next, mount and solder capacitors C1 & C2 (.1uF), C11 (.047uF), and C12 (.01uf). Mount and solder the two 1 Meg, 25-turn potentiometers (R10 & R14). Insert R14 in the correct position and bend flat prior to soldering. Now mount and solder the 5.1V Zener diode (D2), the Audio switch, power jack, headphone jack, and digital output jack. Mount and solder the speaker, transistors Q1, Q3 & Q4 (2N3904), 2-pin headers (P2 & P13) and LED (the longer of the LED terminals is positive (+)) to the PCB. The LED should rise 3/8" from the PCB to the bottom of the LED. This distance will ensure proper placement of the LED when the PCB is mounted inside the case. Mount and solder the bridge rectifier making sure to align the + terminal of the rectifier to the + terminal on the PCB. At this point, your Geiger counter pc board should look like Figure 15.5.

Figure 15.5 Completed Circuit

*R16 is a 10 Meg resistor for LND712 GM Tube (GMT-01).
R16 is 2.2 Megs for GMT-02 GM Tube.
R16 is jumped with wire for external wand.

151

Attaching GM Tube

I will now show you how to connect two different style GM Tubes (the GMT-01 & the GMT-02).

The Geiger-Mueller tube has two leads. The GMT-02* is mounted on the bottom side of the case, see Figure 15.6, below. Use silicon glue to secure tube. The red wire from the GM tube is soldered to +GM tube lead on the PCB. The other wire is soldered to the (-) GM terminal on the PC board.

The Geiger Mueller tube is delicate and needs to be protected in an enclosure. Keep sufficient length of wire so that you can open and close the case.

Figure 15.6. GMT-02 Geiger Mueller Tube in Case

*If using the LND712 Geiger Mueller tube (GMT-01), R16 should be a 10 Meg resistor. The jumper at P10 should be omitted, bringing the output voltage to 550 volts. This Geiger Mueller tube also has two leads. It is mounted on the bottom side of the PCB. The wire connected to the metal sides of the tube is the negative terminal. This is soldered to the (-) GM terminal on the PC board. The center terminal of the GM tube has a removable solder lead. Remove the lead, solder 1.5" of wire to it, and reattach the lead to the center terminal of the GM tube. Take the opposite end of the wire and solder to the (+) GM terminal on the PC board.

The Geiger Mueller tube is delicate and needs to be protected in an enclosure. However, the enclosure has a 1/2" hole that allows the front surface (mica window) of the GM tube to remain exposed. This way, alpha particles can pass through the thin mica window and be detected.

After securing with a wire, as in Figure 15.7, a small amount of glue or epoxy can be dabbed on the wire tube assembly for added support.

Figure 15.7 Detail showing GMT-01 tube secured on PCB

*When attaching an external wand, R16 should be jumped with a small piece of wire. The 10 mega-ohm resistor is located inside the wand with the LND 712 tube. The connector for the wand, Jack-08, is soldered to the bottom of the PC board as shown in Figure 15.8.

Figure 15.8 Detail of Connector Mounted for External Wand

Test before Continuing General Construction

Before mounting the PCB inside the case, check to make sure the entire Geiger counter circuit functions. Background radiation, from natural sources on earth and cosmic rays, will cause the Geiger counter to click. In my corner of the world, I have a background radiation that triggers the counter 12-20 times a minute.

When you are satisfied that the circuit is working properly, we can mount the circuit inside the case. Mount the PC board to the front of the case. The shafts of the two PC mounted switch and LED should fit into the pre-drilled holes. The PCB is held to the case front using the two nuts to the PC mounted switches.

Finish by placing the 9-volt battery cap into the battery compartment of the back case. Close the case and secure with case screws.

Installing /Changing Battery

To install or change battery, open battery compartment on the back of the Geiger counter and install or replace battery, see Figure 15.9.

Figure 15.9

Features:

Headphone

The headphone jack may be used for a headphone. When using the headphone jack for headphones, the speaker is automatically cut-off.

Digital Output and Analog Meter

The digital output provides a TTL logic (+5V) pulse every time the Geiger counter detects radiation. This signal can easily be connected to a microcontroller or PC. The PC or microcontroller can then be used to create a digital Geiger counter, chart recorder, or other recording instrument for nuclear experiments. The digital output may also be used to connect an analog meter to the unit. The analog meter is an accessory that plugs into the digital output and provides a visual indication of the approximate radiation level.

External Power Jack

The GCK-01 may be powered by either a 9-Volt battery or external power source with a 2.5mm jack.

Headphones, power supplies, and other accessories may be found on Images SI website at https://www.imagesco.com

Parts List from Images SI, Inc.

The complete parts list and your choice of tube is available as the GCK-01 kit

(1) PCB-66

(2) 1K 1/4W Resistor - R1, R4

(7) 10K 1/4W Resistors - R2, R5, R6, R7, R8, R12, R19

(1) 15K 1/4W Resistor - R9

(2) 1 Mega Ohm 1/4W Resistor - R3, R11

(1) 100K 1/4W Resistor - R13

(2) 330 Ohm 1/4W Resistors - R15, R21

(1) 10 Mega Ohm 1/4W Resistor - R16 (for GMT-01)

(1) 2.2 Meg Resistor – Alt R16 (for GMT-02)

(1) 5.6K 1/4W Resistor - R17

(1) 4.3K 1/4W Resistor - R18

(1) 470 Ohm 1/4W Resistor - R20

(1) 1 Meg 25-Turn Potentiometers - R14

(3) 0.1uF 50V Capacitors - C1, C2, C10

(3) 0.01uf 1KV Capacitors - C3, C4, C5

(1) 220uF 10V Capacitor - C6

(1) 22uF 50V Capacitor - C7

(2) 0.01uf 12V Capacitors - C8, C12

(1) 0.0047uf 100V Capacitor - C9

(1) 0.047uF Capacitor - C11

(1) 1N751 (5.1V Zener) Diode - D2

(1) Red Subminiature LED - D3

(2) 1N4007 Diodes - D4, D5

(1) 1N5953 (150V) Diode - D6

(2) 1N5956 (200V) Diodes - D7, D8

(1) 1N5817 Diode - D9

(1) 1N914/1N514 Diode - D10

(1) ICS-14

(1) LM339 - U1

(1) W01M - Bridge Rectifier - U2

(1) 7805 Voltage Regulator - U3

(1) ICS-16

(1) 4049 - U4

(1) ICS-8

(1) LM555 Timer - U5

(3) 2N3904 Transistors - Q1, Q3, Q4

(1) IRF830 - Q2

(2) Jack-05

(3) SMH-02

(1) PJ-102A

(1) SPK-05B

(2) SW-07

(1) HVT-03

(1) BAT-01

(1) Jumper

(1) GM Tube (optional)

(1) Plastic Case (Sold separately)

GCK-01 Complete Geiger Counter Kit with multiple GM tube options (all components) at https://www.imagesco.com/geiger/geiger-counter-kits.html

Components that are sold separately (go to links for more info or to purchase):

GM tube (GMT-01) $74.95 https://www.imagesco.com/geiger/geiger-counter-tubes.html

GM tube (GMT-02) $44.95 https://www.imagesco.com/geiger/geiger-counter-tubes.html

Mini Step-up transformer $8.00 https://www.imagesco.com/high-voltage/transformers.html

Radioactive Sources (go to links for more info or to purchase):

Uranium Ore (UR-01) $39.95
https://www.imagesco.com/geiger/uranium-ore.html

CS-137 Source $81.95 - $130.00 * Drop shipped from different
location https://www.imagesco.com/geiger/radioactive-sources.html

Optional LCD Meter

The Analog Digital Radiation Meter counts the pulse output of a
standard analog Geiger counter to provide a visual readout of the CPS,
approximate radiation level (imperial/metric) and analog radiation
field strength meter.

Figure 15.10 Close up of the Analog Radiation Edge Meter.

Features:

- 7-12VDC or 5VDC power supply input
- Counts TTL pulses from Analog Geiger Counter
- Outputs Digital Counts Per Second (CPS) value
- Outputs radiation level (imperial / metric)
- LCD Backlight

Images SI, Inc.'s Analog Digital Meter allows you to add a digital
display to your analog Geiger counter. The default display will provide
readings in CPS and mR/hr. To switch the unit to display metric

measurements (mSv/hr), a jumper is placed on the two-pin header on the back of the PCB.

The standard unit derives its power at either 5VDC or 9VDC, see image below for accurate wiring diagrams.

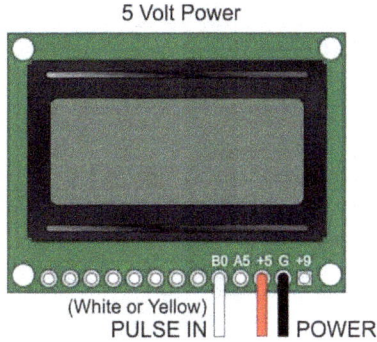

Figure 15.11

The meter features onboard switches that control power to the meter and backlight, as well as a control to adjust the contrast on the LCD. Additional options give you the ability to hardwire the unit so that the power to the meter and backlight remain on at all times while your Geiger counter is on. In other words, "set it and forget it". You may also choose to add an external contrast control.

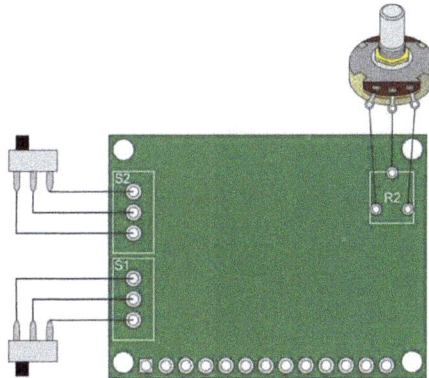

Optional external switches and contrast control.

Figure 15.12

159

More Information at http://www.imagesco.com/geiger/analog-radiation-meter.html

Chapter 16

16 - DMAD Expansion Module

Adding digital capabilities to Analog Geiger Counters

The Digital Meter Adapter is a universal expansion module. This unit combines a Digital Meter Adapter with a Random Number Generator. It also has the capability to connect an analog Geiger counter to a PC using a TTL serial cable. The DMAD adds additional features to enhance the capabilities of Analog Geiger counters.

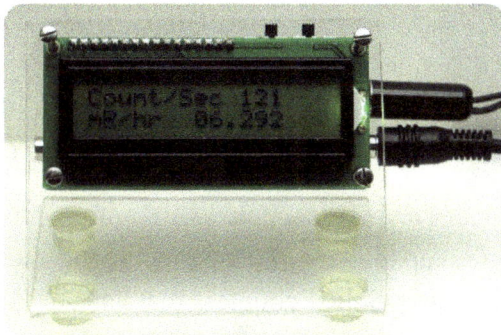

Figure16.1

Notes on DMAD radiation conversion accuracy:

The Digital Meter mode of the DMAD is made to work with a large number of different Geiger counters. While the count in CPS and CPM will be accurate, the conversion of those numbers to a radiation value are only approximate, based upon a general conversion program. They should not be considered accurate.

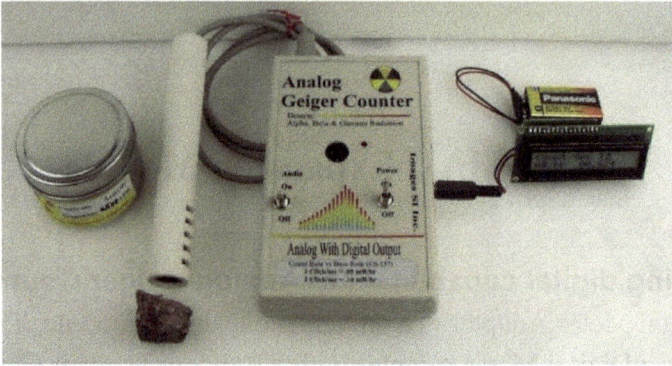

Figure 16.2

The DMAD Adapter

The DMAD allows the user to choose between several different functions. It can currently be used as a Digital Meter with a TTL serial digital output or Random Number Generator. The TTL serial digital output can be used to connect the Geiger counter to a PC. This function requires an additional USB/TTL cable with PC drivers.

The DMAD provides a single module solution for adding various extra features to enhance capabilities of Geiger counters. The DMAD can connect to most Analog Geiger counters that have a +5V TTL pulse output for each radioactive particle detected. To connect to Images Analog Geiger counter, you can use the included male to male 3.5mm mono plug. This plug connects to the digital output of the Analog Geiger counters and the input to the DMAD.

Functionality is selected by a switch located on the backside, A7.

164

Digital Meter Function:

Figure 16.3

Digital Meter provides the option to count in either CPS (Counts per Second) or CPM (Counts Per Minute) on the top line of the LCD display. The bottom line of the LCD display provides the approximate radiation level.

A7 Switch/Up Position = Digital Meter with serial digital output for USB PC Interface

When A7 is not set (switch in up position, open), the module will function as a Digital Meter. The A7 switch is checked for position when the module is turned on, so switch setting should be changed only when module is off. Changing the switch setting while the module is powered will not change functionality of module.

Usage

Switch A7 in up position (Unset) selects Digital Meter mode with digital output, the module has four options that may be selected using the switches on the back of the circuit, see Figure below.

Switch B6 sets whether the output radiation level is given in conventional terms mR/hr (milliRads/milliRem per hour) or System International (SI) mSv/hr (milliSiverts per hour).

165

B7	B6	B5	B4	Function
Up	Up	XXX	XXX	mR/hr
Up	Down	XXX	XXX	mSv/hr
Up	XXX	XXX	Down	60 Seconds
Up	XXX	XXX	Up	1 Second

Up = unset   Down = set   XXX = Don't care

Switch B4 sets the timing mode of either one minute or one second. When B4 is set to one-minute mode, the radiation level is given in µR/hr or µSv/hr. This mode is convenient for checking the local background radiation.

Irrespective of the selection of CPS/CPM, CPS data will be sent out serially every second, which can be interfaced with PC for graphing and charting purposes. The USB/TTL interface cable sends the counts per second information transmitted to the PC via USB in two bytes. A high byte, that is multiplied by 256 and added to the low byte for the total count. The USB/TTL cable is seen on the PC as a COM Port.

To use the DMAD as a Digital Meter, set the switches for the units/time you want, see chart above. Plug the 3.5mm Male to Male cable into the Digital out of the Analog Geiger counter, and the opposite end of the cable into the Serial input (AGC/DGC in), J1 on the DMAD board, see Figures on following page.

Figure 16.5

Turn the Geiger counter on, then turn this module on. After powering on the module, its LCD screen will be empty for 1 second, the module's LCD is initializing during this time. Then the screen will read "**DMAD for AGC**" on the top line and "**Counting...**" on the bottom line of LCD for another 1 second. After that, counts and radiation level calculations will commence as radiation particles are detected by the Analog Geiger counter. The LCD screen will display counts and radiation levels as per time and units selected and radiation particles detected by Analog Geiger counter.

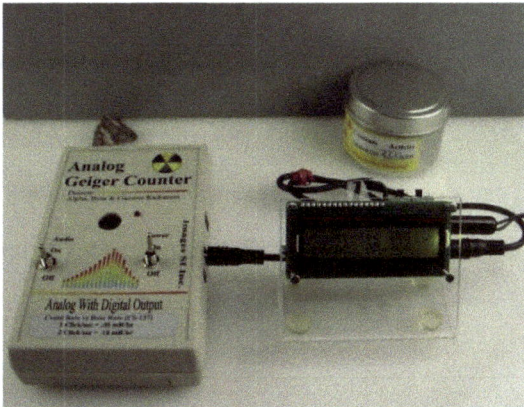

Figure 16.6

If there is no display on the LCD, adjust the contrast control on the back of the Digital Meter Adapter.

USB Adapter

You can use PC's USB port to connect to the DMAD module using USB/TTL Cable. When using the USB to TTL cable, one usually sets the drivers to use either COM port 3 or 7. The graphing program allows you to choose any open COM port available on the PC. You need to choose the COM port that the USB/TTL cable driver is set to.

The USB to TTL cable and WIN PC driver(s) are available on Images SI website at: http://www.imagesco.com/semiconductors/usb-3.5mm.html

USB TTL Serial Cable allows easy interfacing to devices over USB

The USB TTL Cable provides connectivity between USB and serial UART interfaces at 5V.

3.5mm Audio Jack output.
Connector configuration:
tip - TxD
ring - RxD
Sleeve - GND

Windows Geiger Counter Graphing & Logging Software

The DMAD allows you to output your data from the analog Geiger counter to a Windows PC using free Windows graphing software. The amount of data that may be saved is limited by the memory in your system or space available on your hard drive. But it is safe to say you could continue to graph for weeks.

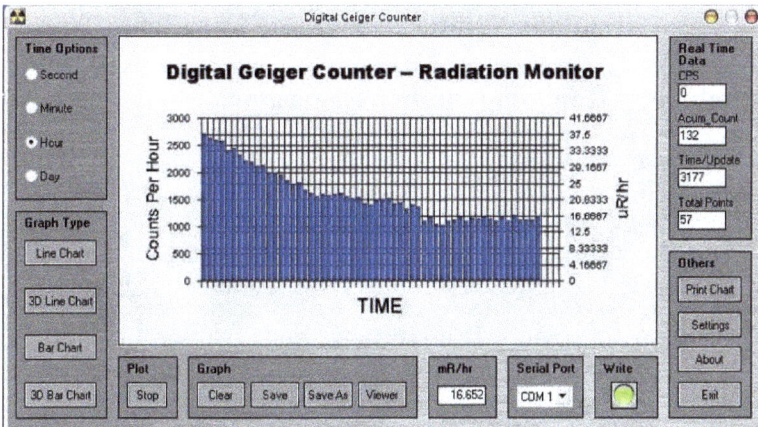

Figure 16.8

Geiger Counter Charting Software is proprietary of Images SI, and is available free of charge at: https://imagesco.com/geiger/geiger-graph.html

Geiger Counter Charting Software runs on Windows. Download the software installation file from Images SI website and install it on your PC. Connect the module to PC's USB port via USB to TTL stereo cable. The module should be set to operate in DMAD / PC interface mode by setting switch A7 Up.

Make sure the program's COM port is set to the correct COM port where the module is connected. Both the module and Geiger counter must be on and properly connected for the program to begin graphing. The graphs generated by the program may be saved to disk and loaded for viewing and analysis later on. See screen image.

169

Random Number Generator Function

A7 Switch Set = Random Number Generator:

Figure 16.9

When A7 switch is in its Down position, it is set, and the module will function as Random Number Generator. Switch A7 is only checked when the module is initially turned on, so switch setting should be changed only when the module is off. Changing the switch setting while the module is powered on will not change functionality of module.

Switch Settings for Random Number Generator				
A7	**B6**	**B5**	**B4**	**Range**
Down	Down	Down	Down	1-2
Down	Down	Down	Up	1-4
Down	Down	Up	Up	1-8
Down	Down	Up	Down	1-16
Down	Up	Up	Down	1-32
Down	Up	Up	Up	1-64
Down	Up	Down	Up	1-128

Random Number Generator functionality generates random numbers true in nature. For this purpose, the module can be used with either Analog or Digital Geiger counters. The module plugs into the Digital

170

out of the Analog Geiger counter or Audio out of the Digital Geiger counter. The range of random numbers generated is selected by the combination of switch settings for B6, B5, B4.

The TTL serial sends the random number generated to the PC via the interface. The information is transmitted as a single byte. This data is used by various PC programs for monitoring and graphing.

When using the DMAD in Random Number Generator mode, the Geiger counter should be set up to read background radiation. Typically, depending upon your location (background radiation is location dependent), the generator will generate 20-40 random numbers per minute. Random numbers generated are truly random since they are based on naturally occurring radioactivity. These random numbers are displayed on the LCD Display and are sent out via a TTL serial port that may be interfaced to a micro controller or PC. Serial communication specifications are: 9600 baud, 8N1.

To use the DMAD as a Random Number Generator, The A7 switch on the back of the module should be set before the module is powered on, in order to select the RNG mode of operation. The module has three option switches, B6 to B4 (see table on previous page), for selecting number range. The current version of RNG module can produce 7 different ranges of random numbers.

Set the switches for the range of your choice. Plug one end of the 3.5mm (1/8") male to male cable into the Digital out of the Analog Geiger counter / Audio out of Digital Geiger counter. Insert the other end of the 3.5mm cable into the J1"Serial in (AGC/DGC in)" socket on the DMAD. Turn the Geiger counter on, then turn the module on. If there is no display on the LCD, adjust the contrast control on the back of the module.

When the module is powered on, the LCD Screen will be blank for 1 second (LCD screen is initialized during this time) and then a splash screen will be displayed with text "**Random Number**" appearing on the top line of the LCD Screen, followed by text "**Generator**" on the bottom line of LCD Screen. The splash screen will be displayed for a minimum of 1 second, after random number generation commences. The module will display the random number generated on the LCD screen only when the Geiger counter detects a radioactive particle. This can take a while (average background radiation particles count per second is 20, if background radiation level is low in your location). Until the first radioactive particle is detected, the LCD Screen will display the splash screen (Figure 16.9).

Once random number generation commences, for each random number generated, the LCD Screen will show text "**Range: 1 to 2**" in top line, (range selected is 1 to 2 for this example). The bottom line will display text "**Generated:**" followed by the random number generated. This display will be present until the next number is generated, after which the display will be refreshed to display current number.

In reality, random numbers generated inside a microcontroller for any selected range starts from 0 instead of 1. Before the generated number is displayed on the LCD Screen and serially sent out, 1 is added to the generated number to shift the range from starting 0 to starting 1. This is done because conventional counting system starts from 1.

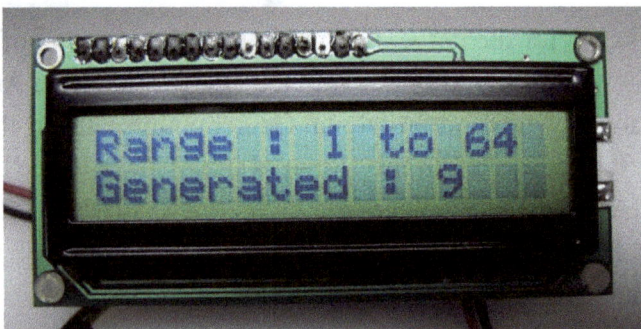

Figure 16.10

As each radioactive particle is detected by the Geiger counter, a new random number is generated and displayed on the LCD. At the same

172

time the number is sent out via the TTL serial port, which can be accessed by header I/O & 3.5mm socket on the back of the DMAD Module. This allows any system to take the random numbers generated by this module for further use. PC can be connected to this module with a USB to TTL cable and a proper interface to read random numbers generated. Serial Data format is 9600 Baud, 8N1 with a byte wide data for random number.

Random Number Generation Theory

True random numbers are useful for data encryption (cryptography), statistical mechanics, probability, gaming, neural networks, and disorder systems, to name a few.

Random numbers generated by a computer are notoriously random in an exact defined pattern. So much so that they are many times referred to as pseudo-random numbers. The random numbers generated by this module are truly random.

The way the RNG module operates is simple to understand. However, its fundamental function is best described using a mechanical analogy. Imagine a sequence of numbers painted on the outer edge of a revolving wheel. There is a pointer on the outside of the wheel that points to the number at the top of the wheel. The wheel is set into motion, spinning very rapidly. Then at any given "random" moment in time, the wheel is instantly stopped, and the number under the pointer becomes our random number. Once read, the wheel is set back into motion again for next random number generation.

The electronics follow pretty close to the mechanical analogy. The microcontroller spins the sequence of numbers internally. The random point in time when the sequence of numbers is instantly stopped is generated by the detection of a radioactive particle by the Geiger counter. Background radiation is an ideal quantum mechanical random time-delay generator because it is impossible to predict with any accuracy the exact moment a radioactive particle will be detected.

When a particle is detected, a positive pulse is sent to the Digital out of Analog Geiger Counter / Audio out of Digital Geiger Counter. This module is connected to Digital out of Analog Geiger Counter / Audio out of Digital Geiger Counter. On RNG Module, the positive pulse is connected to the microcontroller input pin. This positive pulse generates an interrupt in the microcontroller that stops the numbers from spinning. The number is read, displayed, and sent out serially on the TTL port.

We determine how many numbers are painted on the wheel by setting the jumpers on the back of this module i.e. setting the range of random number being generated.

There is a simple test to find the trueness of random numbers. If you add all the random numbers that are generated (Let's call this number N) and divide that by the number of samples you added together (Let's call this number S), the answer should be approximately 50% of your number range (Lets call this number R).

For example, using a range of $1 - 10$, so R = 10. The number of random number samples is 100, so S = 100. We add those 100 random numbers together, that equals N. So, $N/S = .5$ R. In this case, the number N when divided by S (100) should equal approximately 5. This would show an even distribution of random numbers throughout the range of 1 - 10.

Writing Your Own Software to Communicate with the Adapter

You can write your own software to communicate with this module. Serial data is sent out as a two-byte number (most significant byte first) with the following specifications: 9600 Baud, Inverted, 8 data bits, no parity, and one stop bit.

Circuit Construction

Figure 16.12 shows the schematics of the DMAD-04. The circuit uses a pre-programmed 16F88 microcontroller and a standard 16 x 2 LCD display. The module derives its power from either an 9V external battery, or optional power supply. Power is regulated on board to 5V. Input to the module comes via 3.5mm mono jack, J1. The module features a TTL serial out port via a two-pin header or 3.5mm stereo jack, J2, which can be interfaced to a PC using Images' USB/TTL Serial Cable. The back of the DMAD Module has a Power switch, LCD backlight switch, Contrast control, and Selection switches, see Figure 16.13.

 Begin by mounting and soldering the 16-pin header on the bottom of the board (the side with **no** silkscreen) as shown in Figure 16.11.

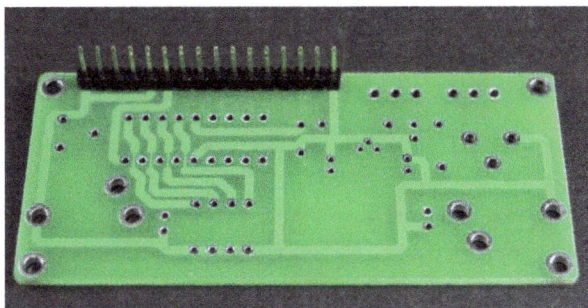

Figure 16.11

Now, on the top of the board (silkscreen side), mount and solder the two slide switches marked S1 & S2, followed by power jack, P3. Next, mount and solder the resistors. R1 is a 33-ohm resistor (color bands orange, orange, black). R4, R5, & R6 are 10K ohm resistors (color bands brown, black, orange). Next mount the 1N4007 diode, D2, making sure to align the stripe on the diode with the strip on the silkscreen.

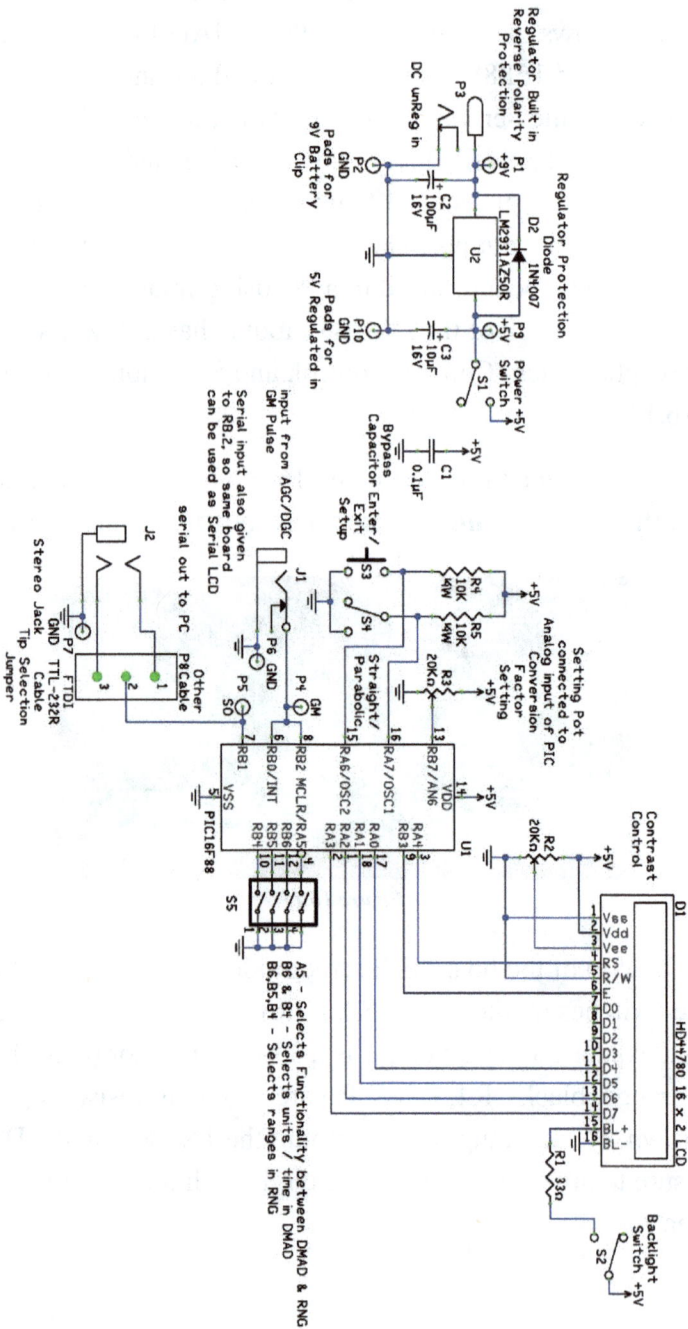

Figure 16.12

176

U2 is the regulator. When mounting the regulator, be sure the flat side is oriented with the flat side of the silkscreen. Next, mount the two 2-pin headers in the top two sets of holes marked "P1, P2" then "P4, P6". Mount the 3-pin header marked P8. These headers provide an alternate method of connecting to the board to supply power, digital input, and serial output respectively.

Now, mount C1, the 0.1uf capacitor, and the two 10uf 16V capacitors marked C2 & C3. When mounting these capacitors, be sure the longer lead is oriented to the hole marked positive.

Mount and solder the Serial Input jack, J1, and Serial Output jack, J2, followed by the 20K pot in the spot marked R2 for the Contrast Control. Mount and solder the 4-position switch in S5.

Next, mount and solder 18 Pin socket U1 to the PC board, making sure that it is oriented according to the outline on the silkscreen. Insert the preprogrammed 16F88 microcontroller matching the notch on the chip to the notch on the socket.

Next, mount and solder the LCD module to the 16-pin header. Figure 16.10 is a photo of the top of the completed circuit.

Figure 16.13

177

DMAD-04 Parts List

(1) PCB-74

(1) PIC16F88   U1

(1) ICS-18

(1) CAP-.1uf-50V C1

(2) CAP-10uf-16V C2 C3

(1) SMH-16   D1

(2) SMH-02   P1&2 P4&6

(1) SMH-03   P8

(1) POT-20K STT R2

(1) RES-33ohm 1/4 R1

(3) RES-10K 1/4 R4 R5 R6

(1) LCD-01-16x2

(1) LDO-5V   U2

(1) SW-26   S5

(2) SW-27   S1 S2

(1) PJ-102B   P3

(1) Jack-05   J1

(1) Jack-13   J2

(1) 1N4007   D2

(1) Male-Male 2 foot, 3.5mm Cable

(1) Power Plug

Optional

Acrylic Stand w/4 machine screws and nuts

USB to TTL Cable

Thank you for reading *Nuclear Experiments Using a Geiger Counter*.

If you enjoyed this book and found it helpful, I would be very grateful if you would post a review via the link below. Your feedback really does make a difference, and it will help me make this book even better for future readers.

Even a one-word comment would be appreciated.

Thank you for your support.

https://www.amazon.com/review/create-review/?ie=UTF8&channel=glance-detail&asin=B07YXDFM3S

Thank you,

John Iovine

www.ingramcontent.com/pod-product-compliance
Lightning Source LLC
Chambersburg PA
CBHW071227210326
41597CB00016B/1970